Know It All!
Grades 6–8 Math

by Diane Perullo

Random House, Inc.
New York

www.randomhouse.com/princetonreview

This workbook was written by The Princeton Review, one of the nation's leaders in test preparation. The Princeton Review helps millions of students every year prepare for standardized assessments of all kinds. The Princeton Review offers the best way to help students excel on standardized tests.

The Princeton Review is not affiliated with Princeton University or Educational Testing Service.

Princeton Review Publishing, L.L.C.
160 Varick Street, 12th Floor
New York, NY 10013

E-mail: textbook@review.com

Published in the United States by Random House, Inc., New York.

ISBN 0-375-76376-7

Editor: Robert Elstein
Development Editor: Scott Bridi
Production Editor: Wade Ostrowski
Director of Production: Iam Williams
Design Director: Tina McMaster
Art Director: Neil McMahon
Production Manager: Mike Rockwitz
Production Coordinator: Robert Kurilla

Manufactured in the United States of America

9 8 7 6 5 4 3

Acknowledgments

I would like to extend my deep appreciation and gratitude to Russell Kahn for asking me to participate in such an innovative project. I would also like to thank my editor, Robert Elstein, for his creative solutions, encouragement, and continuing support. Many others at The Princeton Review are deserving of thanks, including Michael Bernstein, Scott Bridi, Paulo de Freitas Jr., Robert Kurilla, Neil McMahon, and Wade Ostrowski.

This book is dedicated to my darling Annie and her family, Don, Theresa, Jim, and Pete.

Table of Contents

Introduction for Parents and Teachers

About This Book

Know It All! focuses on the math skills that students need the most to succeed in school and on standardized achievement tests while providing information about a wide array of fascinating subjects. The math skills that students will review and practice in this book were chosen based on the math standards developed by the National Council of Teachers of Mathematics.

Know It All! contains chapters covering essential math skills, reviews called Brain Boosters, an answer key for the chapters and Brain Boosters, a practice test, and answers and explanations for the practice test. Each **chapter** focuses on a skill or set of related skills, such as the chapter about fractions, decimals, and percents. Each **Brain Booster** includes practice questions that review the content in the chapters that precede it. The **answer key** provides correct answers to the questions in the chapters and Brain Boosters. The **practice test** is a test that resembles the style, structure, difficulty level, and skills common in actual standardized achievement tests. The **answers and explanations** provide correct answers to the questions on the practice test and explain how students can answer them correctly.

Each **chapter** contains the following:

- an introduction that presents the content covered in the chapter and defines relevant terms
- a sample passage and question followed by a step-by-step explanation of how to answer the question
- passages about interesting subjects and practice questions that cover the content of the chapter
- *Know It All!* tips to assist students in further developing their skills

There will also be cumulative review sections called Brain Boosters that follow every few chapters in the book.

The **practice test** contains the following:

- multiple-choice, short-answer, and open-ended questions similar in wording and difficulty level to questions on actual standardized achievement tests
- a bubble sheet similar to bubble sheets on actual standardized achievement tests for students to fill in their answers to multiple-choice questions

Explanations following the practice test illustrate the best method to solve each question.

About The Princeton Review

The Princeton Review is one of the nation's leaders in test preparation. We prepare more than two million students every year with our courses, books, on-line services, and software programs. We help students around the country on many statewide and national standardized tests in a variety of subjects and grade levels. Additionally, we help students on college-entrance exams such as the SAT-I, SAT-II, and ACT. Our strategies and techniques are unique and, most importantly, successful. Our goal is to reinforce skills that students have been taught in the classroom and to show them how to apply these skills to the format and structure of standardized tests.

About Standardized Achievement Tests

Across the nation, different standardized achievement tests are being used in different locations to assess students. States choose which tests they want to administer, and, often, districts within the state also choose to administer additional tests. Some states administer state-specific tests that assess the state's curriculum. Examples of state-specific tests are the Florida Comprehensive Assessment Test (FCAT) and the Massachusetts Comprehensive Assessment System (MCAS). Other states administer national tests, which are tests used in several states throughout the country. Examples of national tests are the Stanford Achievement Test, Iowa Test of Basic Skills (ITBS), and TerraNova/CTBS (Comprehensive Test of Basic Skills). Some states administer both state-specific tests and national tests.

To find out more information about state-specific tests, go to www.nclb.gov/next/where/statecontacts.html. You can also visit k12.princetonreview.com and click on Assessment Advisor in order to find information about state-specific tests.

Most tests administered to students contain multiple-choice and open-ended questions. Some tests contain short-answer questions. Some tests are timed, and others are not. Some tests are used to determine if a student can be promoted to the next grade, and others are not.

For information about the tests your students will take, the administration and timing of the tests, the effects of the results, or any other information, you should contact your school or your local school district.

No test can assess all of the unique qualities of your student. Standardized achievement tests are intended to show how well a student can apply skills he or she has learned in school to a testing situation.

State-specific tests are connected to a state curriculum, so they show how well students can apply the skills in their curriculum to a testing situation.

National tests are not connected to specific state's curriculum, but they include content that *most likely* would be taught in your student's grade. Therefore, some national tests may assess content that students find unfamiliar because it is not part of the school's curriculum at that grade level. National tests compare a student's performance on the test with the performance of other students in the nation on the same test.

Student Introduction

Student Introduction

About This Book

What kind of person is a *know it all*? Someone who craves information and wants to learn new things. Someone who wants to be amazed by what they learn. Someone who is excited by things that are strange and unusual.

Know It All! is an adventure for your mind. *Know It All!* is chock-full of wild, weird, zany, fascinating, unbelievable, monumental articles—all of which contain true stories! Plan to be stunned, amused, intrigued, and grossed out on this adventure.

In addition to feeding your brain all sorts of interesting information, *Know It All!* will feed your brain with test-taking tips and standardized-test practice.

By the end of this book, you will have the biggest, strongest brain you've ever had! You'll be ready for the Brain Olympics and to run for president of the United Brains of the Universe. You'll be a *know it all*!

Know It All! contains **chapters, Brain Boosters,** and a **practice test.**

Each **chapter**

- defines a skill or group of skills, such as the chapter about fractions, decimals, and percents
- shows how to use the skills to answer a sample question
- provides you with passages and practice questions like those you may see on standardized tests in school
- gives *Know It All!* tips to help you become the *know it all* you want to be

Each **Brain Booster**

- reviews skills in the previous chapters
- includes fun and interesting passages to read
- provides practice questions for you to answer

The **practice test**

- gives you practice answering questions similar to those on standardized tests
- supplies a bubble sheet that is similar to the one you'll see on standardized achievement tests

The book contains an answer key for the questions in the chapters and Brain Boosters. Also, the answers and explanations provide correct answers to the questions on the practice test and explain how you can answer them correctly.

About Standardized Achievement Tests

Standardized achievement tests. Who? What? Where? When? Why? How?

You know about them. You've probably taken them. But you might have a few questions about them. If you want to be a *know it all,* then it would be good for you to know about standardized achievement tests.

The words *standardized* and *achievement* describe the word *tests. Standardize* means to compare something with a standard. Standardized tests often use standards that have been decided by your school, district, or state. These standards list the skills you will learn in different subjects in different grades. *Achievement* means the quality of the work produced by a student, a heroic act, or an impressive result gained through effort. So *standardized achievement tests* are tests that assess the quality of your work with certain skills. According to these definitions, you can consider yourself an impressive hero for all of your effort while taking these tests!

To find out the nitty-gritty about any standardized achievement tests you may take, ask your teachers and parents questions about them. The following are some questions you might want to ask:

Who? You!

What? What kinds of questions will be on the test?

What kinds of skills will be tested by the test?

Where? Where will the test be given?

When? When will the test be given?

How much time will I have to complete the test?

Why? Why am I taking the test?

How? How should I prepare for the test?

Do I need to bring anything to the test?

No test can assess your unique qualities as a *know it all.* The purpose of a standardized achievement test is to show how well you can use the skills that you learned in school in a testing situation.

Things to Remember When Preparing for Tests

There are lots of things you can do to prepare for standardized achievement tests. Here are a few examples.

- **Work hard in school all year.** Working hard in school throughout the year is a great way to prepare for tests.

- **Read.** Read everything you can. Read your homework, your textbooks, the newspaper, magazines, novels, plays, poems, and comics—even the back of the cereal box. Reading a lot is a great way to prepare for tests.

- **Work on this book!** This book provides you with loads of practice for tests. You've probably heard the saying "Practice makes perfect." Practice can be a great way to prepare for tests.

- **Ask your teachers and parents questions about your schoolwork whenever you need to.** Your teachers and your parents can help you with your schoolwork. Asking for help when you need it is a great way to prepare for tests and become a *know it all.*

- **Ask your teachers and parents for information about the tests.** If you have questions about the tests, ask them! Being informed is a great way to prepare for tests.

- **Have a good dinner and a good breakfast before you take a test.** Eating well will fuel your body with energy, and your brain thrives on energy. You want to take each test with all the engines in your brain running properly.

- **Get enough sleep before you take a test.** Being awake and alert while taking tests is very important. Your body and your mind work best when you've had enough sleep. So get some Zs on the nights before the tests!

- **Check your work.** When taking a test, you may end up with extra time. You could spend that extra time twiddling your toes or timing how long you can go without blinking. But if you use extra time to check your work, you might spot some mistakes—and improve your score.

- **Stay focused.** You may find that your mind wanders away from the test once in a while. Don't worry—it happens. Just say to your brain, "Brain, it's great that you are so curious, imaginative, and energetic. But I need to focus on the test now." Your brain will thank you later.

About the Icons in This Book

This book contains many different small pictures, called icons. The icons tell you about the topics of the articles in the book.

Alternative Animals Read these passages to learn about animals that you never knew existed and feats that you never knew animals could accomplish. You'll learn about the biggest, smallest, oldest, fastest, and most interesting animals on the planet. You can find these passages in chapters 2, 3, 6, and 7 and in Brain Booster 3.

Hip History Your mission is to storm some of the coolest castles in history with extraordinary historical figures—some of whom aren't much older than you. These passages will help you complete that mission while learning some of the most interesting stories in history. You can find these passages in chapters 1, 4, 6, 11, 12, and 14.

For Your Amusement You want to play games? Read these passages to learn about cool games, toys, amusement parks, and festivals. You can find these passages in chapters 5, 7, and 10.

Extreme Sports Read these passages to learn about outrageous contests, wacky personalities, and incredible feats in the world of sports. You may not have even heard of some of these sports! You can find these passages in chapters 1, 3, 11, 16, 17, and 18.

Grosser Than Gross How gross can you get? Read these passages if you want to learn about really gross things. Be warned: Some of the passages may be so gross that they're downright scary. You can find these passages in chapters 4 and 20 and in Brain Booster 2.

Mad Science If you read these passages, you'll see science like you've never seen it before. You'll learn about all sorts of interesting science-related stuff. You can find these passages in chapters 17 and 18.

Outer Space Oddities Do you ever wonder what goes on in the universe away from planet Earth? Satisfy your curiosity by reading these passages about astronomical oddities. You can find these passages in chapters 11, 16, and 19.

Explorers and Adventurers Did you ever want to take a journey to learn more about a place? Well, you'll get the opportunity to do that if you read these passages about explorers and adventurers. You can find these passages in chapters 2, 9, and 15 and in Brain Booster 1.

The Entertainment Center Do you enjoy listening to music or watching television and movies? Well, here's your chance to read about them! You can find these passages in chapters 2, 10, and 19 and in Brain Booster 1.

Art-rageous Are you feeling a bit creative? Read these passages to get an unusual look at art that's all around you: books, drawings, paintings, and much more. You can find these passages in chapters 5, 8, 12, 13, and 15.

Bizarre Human Feats People do some very strange stuff. You can read about some of these incredible-but-true deeds in these passages. You can find these passages in chapters 8, 9, 13, and 20.

Wild Cards You'll never know what you're going to get with these passages. It's a mixed bag. Anything goes! You can find these passages in chapters 1, 4, and 14.

The Chapters

Question Types

What was the tallest structure in the world for more than 4,000 years?

Where is one of the longest suspension bridges in the world?

In what sport do mushers compete?

Question Types

Tests can have all sorts of different types of questions. There are multiple-choice, short-answer, and open-response questions. Each type of question has its own quirks. If you follow the Know It All Approach, you'll be all set, and you'll be able to answer each one!

Multiple-Choice Questions

Multiple-choice questions offer several answer choices. After you read a multiple-choice question, look at the answer choices and then choose the answer you think is correct. The best part about multiple-choice questions is that the correct answer is always one of the answer choices; it's sitting right there on the page, ready to bite your nose!

Here's the Know It All Approach for answering multiple-choice questions.

Step 1 **Read the question carefully.**

Ah, yes, the important first step in answering any test question. When you read the question carefully, you can determine exactly what you have to do to answer the question.

Step 2 **Read the question a second time and notice the words and information you need to figure out the answer.**

Some tests may CAPITALIZE, *italicize,* or **bold** important words such as *not, least,* and *estimate.* But even if the test does not do this for you, you should notice those words as well as numbers and other words that will help you solve the problem.

Step 3 **Calculate the answer.**

When the question you are trying to answer requires calculations, you should write clearly so that you can read what you've written. Take a moment to compare what you've written to the information in the test question, just to make sure that you copied everything correctly. Then, calculate the answer by working carefully.

Step 4 **Double-check your answer.**

There are several ways to double-check your answer. For example, you could use addition to check a subtraction problem. Here are some other ways to check your answers.

- Review the information in the question and compare it to the information you used in your calculations.
- Make sure that your answer is reasonable. If you were asked to add 12, 8, and 17 and your answer was 162, something is fishy. The answer 162 is not reasonable, because it is much too large. You can figure that out by looking at the three numbers you added.
- Try rounding the numbers in the question to get a quick estimate of the answer. If you use the numbers above, you would round 12 to 10, 8 to 10, and 17 to 20. $10 + 10 + 20 = 40$. Your actual answer should be around 40.

Step 5 **Read all of the answer choices.**

You may think that you have found the correct answer before reading through all of the answer choices. But it is best to read all of the answer choices in order to **be sure** that you have found the correct answer choice. Sometimes things just slip through the cracks. You may notice something about a problem by looking at the answer choices after you think you've solved it.

Step 6 **For questions that you find difficult, use Process of Elimination (POE).**

Make it easier to find the correct answer by drawing a line through answer choices that you *know* are incorrect. Get rid of them so that you will know that you've already considered them. If you use Process of Elimination (POE), you will have a better chance of finding the correct answer to a difficult multiple-choice question. How? Well, even if you can't calculate the right answer, you may be able to identify *incorrect answers*. Maybe an answer choice is way too unreasonable or just plain silly. If you find that an answer choice is incorrect, cross it out if you are allowed to write on your test booklet.

You may be able to get rid of all of the answer choices except one. If that happens, there's a good chance that it is the correct answer!

You may be able to get rid of one or two of the answer choices. Then, you will have to make an educated guess from the remaining answer choices. Eliminating even one answer choice increases your chances of picking the correct answer.

Step 7 **Note the correct answer choice and fill in the corresponding bubble.**

After you decide which answer choice is correct, carefully fill in the bubble that goes with it. When you fill in the bubble, be sure that you're filling in a bubble for the correct question number.

Directions: Try the following multiple-choice question. Make sure to read the passage below first.

That's a Great Pyramid!

The Great Pyramid of Egypt at Giza was the tallest structure in the world for more than 4,400 years, until the Eiffel Tower was built. As the name suggests, the Great Pyramid is exceptionally large. It was built so well that it still exists today almost entirely intact. The pyramid is composed of approximately 2.3 million blocks of stone. Each gigantic block weighs more than 30 average adults put together.

► The estimated weight of the Great Pyramid is 5,750,000 tons. How would this number be written in scientific notation?

A 5.75×10^4
B 5.75×10^5
C 5.75×10^6
D 5.75×10^7

Follow the steps to find the correct answer to this question.

Step 1 **Read the question carefully.**

The question states, "The estimated weight of the Great Pyramid is 5,750,000 tons. How would this number be written in scientific notation?"

Step 2 **After reading the question once, read it again and notice the words and numbers you need to figure out the answer.**

In this question, the information you'll need to answer the question is the number 5,750,000 and the term "scientific notation."

Step 3 **Calculate the answer.**

Write the number 5,750,000 on a piece of paper. The answer choices have the decimal after the first 5. Place a decimal in the same position on your number. Now, count from the decimal to the last zero to determine which power of 10 is correct. There are six places, so answer choice (C) is correct.

Step 4 **Double-check your answer.**

Make sure you've copied the number correctly from the question. Then, count the number of places from the decimal to the last zero. In this case, the exact power was listed in one of the answer choices.

Step 5 **Read all of the answer choices.**

Perhaps you missed something the first time. What if you misread one of the answer choices? It's always best to keep on the safe side.

Step 6 **Even though you know the correct answer, you can still use Process of Elimination to check your work.**

Eliminate choice (A) because $5.75 \times 10^4 = 57{,}500$, not 5,750,000. Eliminate choice (B) because $5.75 \times 10^5 = 575{,}000$, not 5,750,000. Finally, eliminate choice (D) because $5.75 \times 10^7 = 57{,}500{,}000$, not 5,750,000. It's too much.

Step 7 **Note the correct answer choice and fill in the corresponding bubble.**

Short-Answer Questions

Short-answer questions ask you to answer a question in a short written response. In most short-answer math questions, the answer will be a number, mathematical expression, or phrase.

Here's the Know It All Approach for answering short-answer questions.

Step 1 **Read the question carefully.**

This is always the first step in answering any question on any test. If you take the time to do this, you can avoid making careless mistakes.

Step 2 **Notice the words and information you need to figure out the answer.**

It may help you to write down words such as *each, total, greatest, actual,* and *approximately* on scratch paper.

Step 3 **Calculate the answer.**

Write the information that you need to calculate the answer. Make sure that you've written numbers and symbols *exactly* as they appear in the question. When you calculate, work carefully to prevent mistakes.

Step 4 **Double-check your answer.**

There are several ways to double-check your answer. For example, you could use addition to check your answer to a subtraction problem. You could also review the information in the question and compare it to the information you used in your calculations.

Step 5 **Write your final answer in the space provided.**

Short-answer questions are what they sound like: questions that require answers that are short. That means you won't have much space to write your answer. Most of the time, the answer is a number, mathematical expression, or phrase.

Directions: Try the following short-answer question. Make sure to read the passage below first.

Bridge over Troubled Water

When the Akashi-Kaikyo Bridge in Japan opened in April 1998, it was the longest suspension bridge in the world. Connecting Kobe, in the mainland, with Awaji Island, the bridge is 3,910 meters in length, more than 2 miles, or about the same as 30 football fields long!

Throughout the year, the coast of Japan experiences some very dangerous weather. As a result, the bridge is resistant to both wind and earthquakes. The bridge was built to withstand massive earthquakes reaching 8.5 on the Richter scale as well as hurricane-speed winds of up to 80 meters per second.

▶ Kiko was visiting her family in Japan and wanted to drive across the Akashi-Kaikyo Bridge. Unfortunately, Kiko's car broke down exactly $\frac{1}{3}$ of the way across the bridge. To the nearest meter, how far did Kiko travel on the bridge? Remember that the Akashi-Kaikyo Bridge is 3,910 meters in length.

____

Go through the step-by-step process to find the answer.

Step 1 **Read the question carefully.**

This question asks you to find the length of one-third the distance across the Akashi-Kaikyo Bridge. In this case, you'll need to multiply by $\frac{1}{3}$, which is the same as dividing by 3.

Step 2 **Notice the words and information you need to figure out the answer.**

The information you need to find the answer is "exactly $\frac{1}{3}$ of the way," "to the nearest meter," and "3,910 meters in length."

Step 3 **Calculate the answer.**

You need to find $\frac{1}{3}$ of 3,910 meters. Divide 3,910 by 3 and you will get $1{,}303.33\overline{3}$. To the nearest meter, that is 1,303 meters.

Step 4 **Double-check your answer.**

Check to see that you're using the correct numbers. Then, you can multiply 1,303 by 3 and get 3,909, which is approximately 3,910. Your answer is correct.

Step 5 **Write your final answer in the space provided.**

Make sure that you write the answer the question has asked for. In this case, the question has specified that you're to round to the nearest meter. The correct answer is 1,303 meters.

Open-Response Questions

Open-response questions ask you to write your answer to a question and show your work. Sometimes these questions include more than one part for you to answer.

Here's the Know It All Approach for answering open-response questions.

Step 1 **Read the question carefully.**

Bet you know that already! It's a very important part of taking a test, because reading carefully helps figure out what each question is asking. For instance, you might find a question that asks for an estimate. To receive full credit, your answer *must* be an estimate, *not* an exact number.

Step 2 **Read the question again and identify the words and information that you need in order to calculate the answer.**

It may help you to notice words such as *not, each, total, least, greatest, actual, approximately,* and *estimate.*

Step 3 **Use the space provided to calculate the answer.**

Write neatly and show all of your work. You should always show all of your work when answering open-response questions. If you aren't able to get to the correct answer, you may receive partial credit for showing your work and trying to answer it. Be sure to write neatly.

Step 4 **Double-check your answer.**

There are several ways to double-check your answer. For example, you could use addition to check your answer to a subtraction problem. You could also review the information in the question and compare it to the information you used in your calculations.

Step 5 **Make sure you answer all parts of the question.**

Write the answer or answers to the question very clearly. Some open-response questions include more than one part for you to answer. Answer each part of the question clearly.

Directions: Try the following open-response question. Make sure to read the passage below first.

A Truly Amazing Race

There's an amazing race that lasts for more than 1,150 miles through a difficult but beautiful landscape. It's called the Iditarod Trail Sled Dog Race. In the past, the trail has run through jagged mountain ranges, alongside a frozen river, through dense forest, and over deserted tundra. The mushers and their dogs must cross miles of windswept coast in temperatures far below zero with winds that can cut visibility down to nothing. Around 65 teams start the race in Anchorage, Alaska, though many do not make it to the finish line in Nome.

Part A

Doug Swingley won the Iditarod Race in 1995. He completed the 1,150-mile course in 9 days. Write an algebraic expression that shows an average of how far the winning sled traveled each day of the race.

____

Part B

Using the expression you wrote for Part A, to the nearest mile, approximately how far did the winning sled travel each day?

Show your work.

____

Now, solve this problem using the step-by-step process.

Step 1 **Read the question carefully.**

In this case, you are given a two-part question. You have to write an algebraic equation that shows the average distance the winning sled traveled each day. After you write the correct equation in Part A, you solve it in Part B.

Step 2 **Read the question again and identify the words and information that you need in order to calculate the answer.**

You should write "1,150-mile," "9 days," "algebraic expression," and "each day" on a separate sheet of paper.

Step 3 **Use the space provided to calculate the answer.**

In Part A, there is no calculation; you would write the equation $9x = 1{,}150$. For Part B, you would write your calculations as follows:

$$9x = 1{,}150$$

$$x = \frac{1{,}150}{9}$$

$$x \approx 128 \text{ miles}$$

Step 4 **Double-check your answer.**

Look again at what you underlined in Part A and make sure you wrote the correct equation. Then, look at Part B and make sure you solved it correctly. You can multiply 128 by 9 and get 1,152 to know that you are correct.

Step 5 **Make sure you answer all parts of the question.**

You would want to make sure that you answered all parts of the question and that you had given the answer in the form required. Part B asked for you to round to the nearest mile. If you had answered 127.78 miles, you would have lost some credit.

Subject Review

In chapter 1, you learned about multiple-choice, short-answer, and open-response questions. Remember, the best way to answer each type of question is to use the steps illustrated in the Know It All Approach.

Now, here are the answers from the questions at the beginning of the chapter.

What was the tallest structure in the world for more than 4,000 years?

The Great Pyramid of Egypt at Giza was the tallest structure in the world until the Eiffel Tower was completed in Paris in 1889.

Where is one of the longest suspension bridges in the world?

The Akashi-Kaikyo Bridge was the longest suspension bridge in the world when it opened in 1998. It connects Kobe with Awaji Island in Japan.

In what sport do mushers compete?

Mushers are the navigators of a sled drawn by dogs, and they compete in dogsled races. Strange name, huh?

CHAPTER 2

Rational Numbers

What Broadway musical takes the audience to the African savannah?

What exactly is the Bermuda Triangle anyway?

What happens when a salamander loses its tail?

Different Types of Numbers

You may have seen the terms *rational numbers* and *irrational numbers* before, but take a minute to read the definitions below.

Rational numbers are numbers that can be expressed as a fraction, $\frac{a}{b}$, when a and b are integers and b is not equal to 0.

Irrational numbers are numbers that cannot be expressed as $\frac{a}{b}$, when a and b are integers and b is not equal to 0. Irrational numbers are usually expressed as decimals that continue infinitely without repeating.

Those are the formal definitions. Look at some examples to help you to form a clearer understanding.

Rational numbers: 176, $\sqrt{9}$, –29, 0.8, $\frac{12}{25}$, $-\frac{4}{9}$, $6.4\overline{5}$

Irrational numbers: $\sqrt{5}$, π, 0.131131113 . . ., $\sqrt{2}$, –8.524711769 . . .

If you use a calculator to find the square root of 5 or 2, you'll get a decimal that doesn't repeat or end. That means those values are irrational, because irrational numbers do not repeat or end and cannot be expressed as fractions.

Estimating

Estimating is a way to find an approximate number. When you estimate, you find a number that is close to the exact answer but is not as precise.

If you were having a Halloween party and were going to serve spider cookies—mmm, spiders—you would need to estimate how many cookies you'd need. You might figure that you'd need about 3 for each person because some people would eat more—like your Uncle Frank—and others might eat less. If you were expecting 10 people to show up, you might want about 30 cookies. Or maybe you'd make it an even three dozen, just in case someone was really, really hungry—like Uncle Frank.

Sometimes an estimate just won't do; you need an exact answer for many problems. Say that in addition to spider cookies you were also providing a gift bag of goodies for each person at your party. The gift bags could be quite expensive, and you wouldn't want to make too many. In that case, you would want *exactly* one for each person, or *exactly* 10.

Directions: Read the passage below and answer the question that follows.

The King of Broadway

Perhaps the most amazing part of the Broadway musical *The Lion King* is the dazzling, colorful puppets and costumes designed by director Julie Taymor. Audience members are transported to the African savannah and treated to the wondrous sights of towering giraffes, darting birds, monstrous elephants, and gliding antelopes, all represented with bursting color.

▶ A performance of *The Lion King* is showing in a theater with 45 rows of seats. If there are about 32 seats in each row, what is an estimate for the number of seats in the theater?

A 77
B 1,000
C 1,500
D 14,400

Know It All Approach

Read the question and notice the information you'll need to answer it. The words "estimate for the number of seats" and the numbers 45 and 32 are the important information.

To calculate the answer, you will need to estimate. In order to estimate, you need to round each number you are multiplying. To determine the number of seats in the theater, you should round 32 down to the nearest ten, which is 30, and round 45 up to the nearest ten, which is 50. Because $30 \times 50 = 1{,}500$, your estimate is 1,500 seats in the theater.

Now, you need to double-check your answer. A great way to check your answer to a multiplication problem is to divide. If you divide 1,500 by 30, you should get an answer of 50. Because 50 is your estimate for the number of rows of seats in the theater, you'll know you have calculated correctly.

Use Process of Elimination to eliminate incorrect answer choices. Even if the number you calculated is not 1,500, you can eliminate some choices you know are incorrect. You can eliminate answer choice (A), 77, because it is way too small. You can eliminate choice (D), 14,400, because it is too large. Choice (C) is correct, however. Even if you didn't round 32 down to 30 and 45 up to 50, if you multiply 32 by 45, you will receive an answer of 1,440, which is the closest number to 1,500 of the answer choices.

Don't forget to note the correct answer choice.

Directions: Read the passage below and answer the questions that follow.

Could It Be Sea Monsters?

There is an area off the coast of Florida called the Bermuda Triangle, where ships and planes have occasionally disappeared without a trace. Over the years, strange, unidentified lights have been reported, radio communications have been interrupted, and navigational tools have become unreliable. But what caused these odd phenomena? Theories range from sea monsters, space aliens, magnetic whirlpools, and time warps to new dimensions! There has never been a definite explanation for the strange events. In many cases, it's as if the vessels simply vanished into thin air, becoming the basis of a myth that encourages our fascination with the unknown.

1. In 1909, the world-famous explorer Joshua Slocum disappeared at sea on his way to South America. Slocum was famous because in 1898 he became the first person to sail around the world alone. The 46,000-mile trip took him 3 years to complete. To the nearest thousand miles, estimate how far Joshua Slocum traveled on average each year.

   A 15 miles
   B 150 miles
   C 1,500 miles
   D 15,000 miles

2. Some people blame bad weather for the disappearances in the Bermuda Triangle. The whirling winds of a waterspout, also called a wet tornado, can reach speeds of 150 miles per hour. The powerful waves of a tsunami can travel at speeds more than 500 miles per hour. About how many times faster is a tsunami than the winds of a waterspout?

   A 2 times faster
   B 4 times faster
   C 8 times faster
   D 16 times faster

3. In December 1872, a ship called the *Mary Celeste* was found drifting just outside of the Bermuda Triangle. The *Mary Celeste* had no one aboard. Money, food, and cargo were all found untouched, but the entire crew was gone. The *Mary Celeste* measured 103 feet long and weighed 282 tons. Estimate to find the approximate number of tons per foot of the ship.

   ____

4. **Part A**

One mysterious disappearance in the Bermuda Triangle was a huge ship called the U.S.S. *Cyclops*. This ship was a U.S. Navy ship that transported coal and was estimated to be 540 feet long and weigh 19,000 tons. Use your calculator to find the number of tons per foot of this ship.

____

Part B

Is your answer a rational or an irrational number?

____

Directions: Read the passage below and answer the questions that follow.

Lions, Cheetahs, and Crocodiles, Oh My!

Many animals hunt other animals for food. Cheetahs streak at up to 65 miles per hour to overtake slower animals. Lions spring from the ground to attack unsuspecting prey. Crocodiles hide beneath the water until unwary animals can be grasped in their jaws. But animals that are prey to bigger, faster, or stronger predators have ways to protect themselves. If they didn't, they would quickly become extinct.

5. A mole uses its front feet to dig underground tunnels. It is able to dig a tunnel at 12 feet per hour. How many feet of tunnel would a mole be able to dig in $7\frac{1}{2}$ hours?

   A 85 feet
   B 90 feet
   C 95 feet
   D 100 feet

6. A salamander may detach its tail if it gets caught by a predator. If a salamander's tail falls off or is bitten off, the salamander can regrow the tail. If it takes 4 weeks to grow a 6-inch tail, how much does the tail grow per week?

   A 1 inch
   B 1.2 inches
   C 1.33 inches
   D 1.5 inches

7. A 1.5-meter-long kangaroo can clear 9 meters in a single bound. How many times the length of the kangaroo is a leap of 9 meters?

   ____

8. **Part A**

   A basilisk lizard can escape danger by dropping from a tree onto water and running on the surface of the water for a distance before sinking. The lizard is able to reach a speed of 22 feet per second. If the lizard runs 9 feet at 22 feet per second, how many seconds will it be running on water?

   Show your work.

   Part B

   Is your answer a rational or an irrational number?

   ____

9. A pangolin, or scaly anteater, may grow to 6 feet in length and have a giant tongue that is 23 inches long. Estimate how many times longer the pangolin's body is than its tongue.

____

Subject Review

In chapter 2, you worked with rational and irrational numbers. Remember that an irrational number is one that cannot be expressed as $\frac{a}{b}$, when a and b are integers and b is not equal to 0. They are usually expressed as decimals that never repeat. All other numbers are rational.

You also estimated some answers, finding the approximate number instead of an exact number. You estimate when you need to find quickly a number that is close to the exact number. Make sure that when you estimate, you follow the directions about what type of number is required for the answer; the question could ask for a whole number or a decimal.

Remember the questions from page 27? Well, here are the answers.

What Broadway musical takes the audience to the African savannah?

The Lion King, *which was originally an animated film, later became one of the most popular shows on Broadway.*

What exactly is the Bermuda Triangle anyway?

The Bermuda Triangle is the geographic location between Miami, Puerto Rico, and the island of Bermuda, where many ships and planes have disappeared.

What happens when a salamander loses its tail?

A salamander's tail grows back after it intentionally discards it as a means of fleeing a predator.

CHAPTER 3

Fractions, Decimals, and Percents

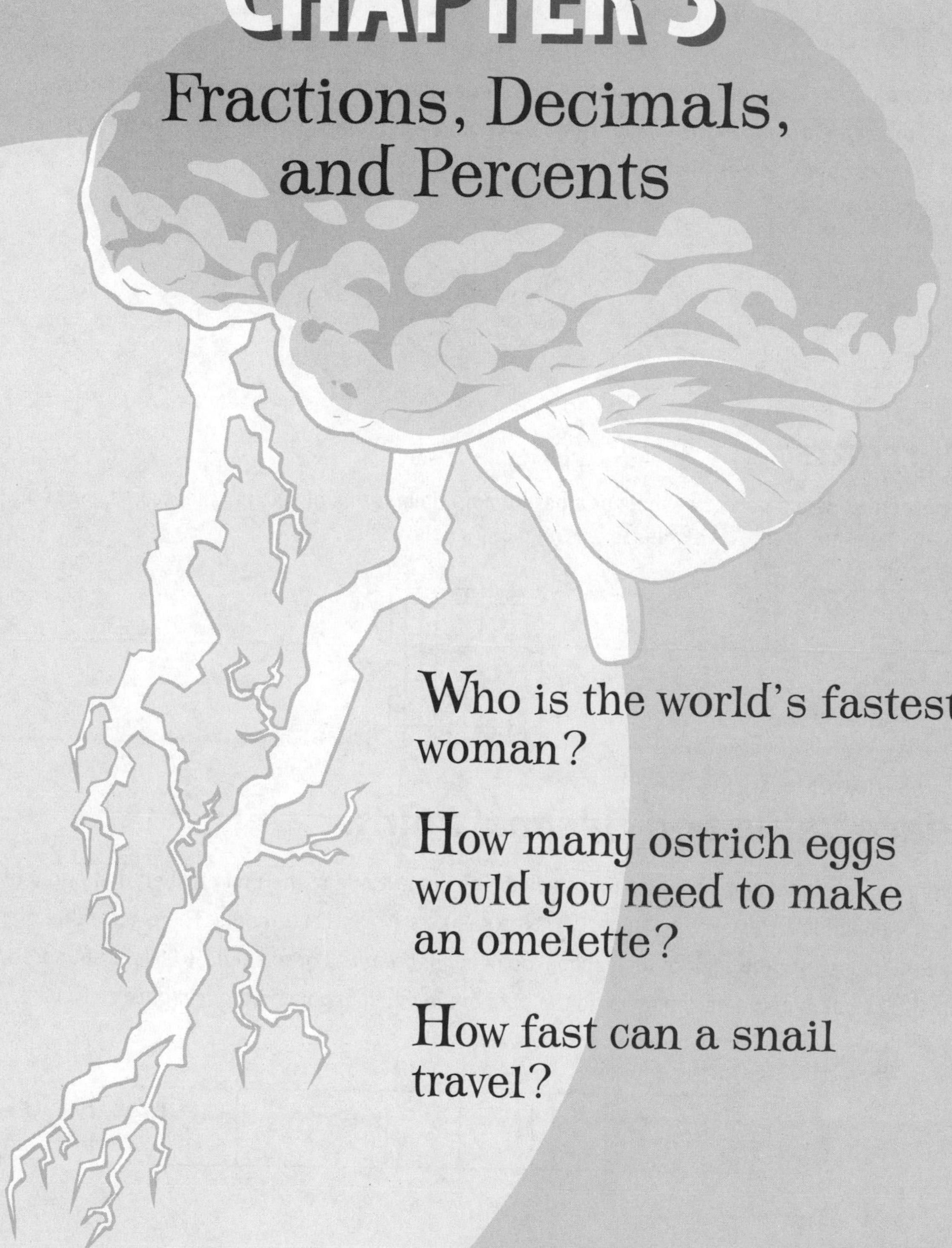

Who is the world's fastest woman?

How many ostrich eggs would you need to make an omelette?

How fast can a snail travel?

Fractions

A **fraction** shows a part of a whole or a group. For example, the picture below shows three parts out of eight parts in the whole. Three parts out of eight parts is written as the fraction $\frac{3}{8}$. The picture below shows the fraction $\frac{3}{8}$.

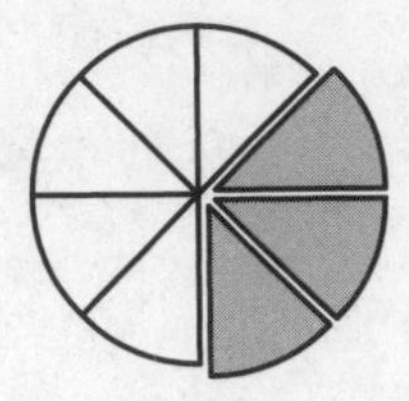

Percents

A **percent** shows a part of a whole or a part of a group in terms of 100. In other words, percent means "per hundred." Its symbol is %. The picture below shows 60%.

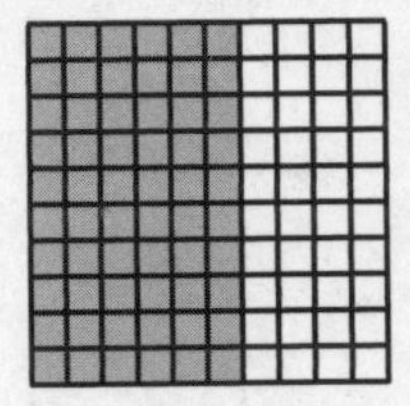

Place Value and Decimal Points

Every digit within a number has a place value. The place value of the first number to the right of the decimal point is *tenths.* For instance, 0.3 is the same as $\frac{3}{10}$, or three-tenths. The place value of the second number to the right of the decimal point is *hundredths.* For example, 0.06 is six-hundredths. The place value of the third number to the right of the decimal point is *thousandths,* so 0.008 is eight-thousandths. The picture below shows 0.3.

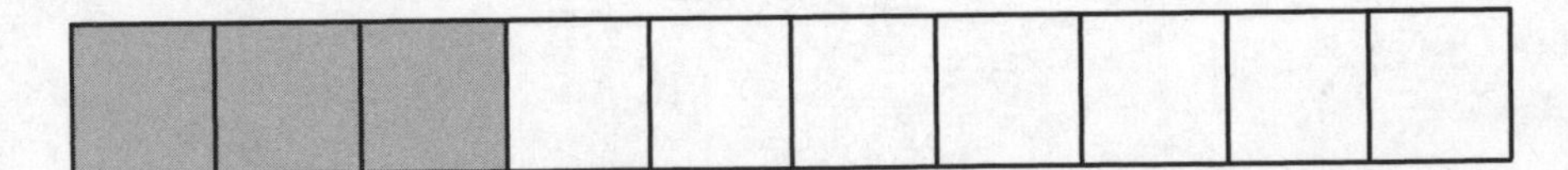

Converting Numbers

You need to know how to convert percents and decimals, especially when you want to compare or order numbers. To convert a decimal to a percent, multiply the decimal by 100 and add the percent sign. (Multiplying by 100 is the same as moving the decimal point two places to the right.)

$$0.45 \times 100 = 45\%$$

To convert a percent to a decimal, divide the percent by 100 and remove the percent symbol. (Dividing by 100 is the same as moving the decimal point two places to the left.)

$$74\% \div 100 = 0.74$$

It's also valuable to learn how to convert between fractions and decimals. Take a look at the fraction $\frac{1}{8}$. To convert a fraction to a decimal, you need to remember that $\frac{1}{8}$ means $1 \div 8$. If you write $1 \div 8$ in long division form and divide, you will get the decimal 0.125.

Converting decimals to fractions is even simpler. To convert 0.4 to a fraction, use the 4 as the numerator and 10 as the denominator. $0.4 = \frac{4}{10}$, which is $\frac{2}{5}$ when simplified. Another example is 0.17, which becomes $\frac{17}{100}$ (remember that the decimal ends in the hundredths place). And 0.251 becomes $\frac{251}{1000}$.

Comparing and Ordering Numbers

Comparing and ordering fractions, decimals, and percents can be difficult. Take a look at $\frac{7}{8}$, 0.65, and 81%. What is their order from smallest to largest? It's not immediately clear. The best way to order these values is to convert them into the same form, such as decimals. Well, 0.65 is already a decimal, so you don't need to change it. But $\frac{7}{8}$ and 81% are not decimals. You can change 81% to the decimal 0.81, because $\frac{81}{100} = 0.81$. But what is the decimal value of $\frac{7}{8}$? Divide 7 by 8 and you will get 0.875. So, the order of these numbers from smallest to largest is 0.65, 81%, $\frac{7}{8}$.

Another way of ordering numbers is by looking at visuals. Take the numbers $\frac{13}{20}$ and $\frac{2}{5}$. Which is larger? Fraction strips can help.

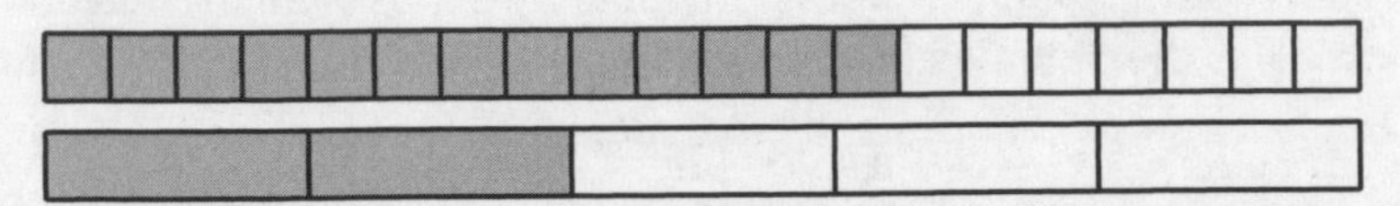

You can see by the shaded portion of the fraction strips that $\frac{13}{20}$ is larger than $\frac{2}{5}$.

Directions: Read the passage below and answer the question that follows.

Hold the Bacon

Like birds, bird eggs come in a variety of sizes. The huge ostrich has a huge egg, usually around 21 centimeters long and 15 centimeters wide. A chicken has a much smaller egg, usually about $5\frac{3}{4}$ centimeters long and $4\frac{2}{5}$ centimeters wide. Some eggs are teeny-tiny, like those of the little hummingbird. Their eggs are only about 1 centimeter long by 1 centimeter wide. Thousands of them can fit into a single ostrich egg. Can you imagine how many it would take to fill the egg of a *T. rex*?

▶ On the number line below, mark and label the following egg widths (in centimeters): 2.3, labeled T for turtledove; $1\frac{1}{2}$, labeled R for robin; 3.3, labeled V for raven; and $1\frac{3}{4}$, labeled P for purple martin.

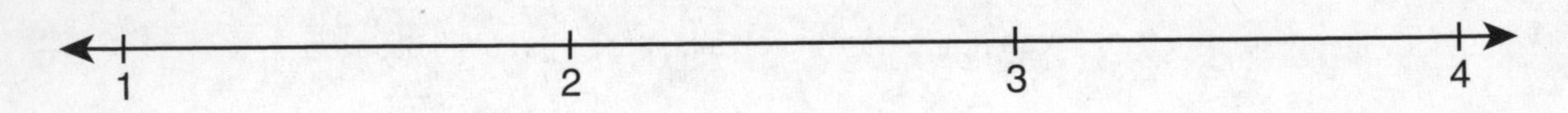

Know It All Approach

After reading the problem carefully, notice the numbers and labels 2.3, T, $1\frac{1}{2}$, R, 3.3, V, $1\frac{3}{4}$, and P. Instead of calculating an answer, you will be placing marks on the number line with letters above them to show the points that the problem has required. It is essential to be neat when placing your marks and your letters!

Take each number and letter in order and place them on the number line. Your finished answer should look like the number line below.

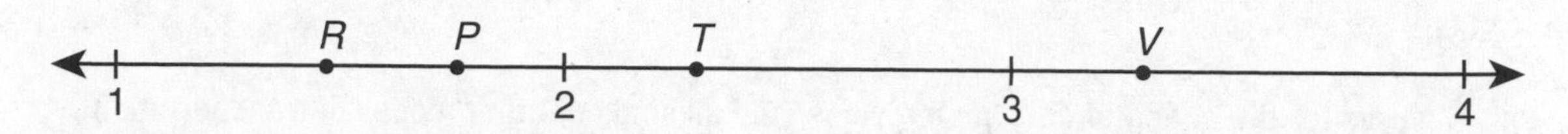

Remember to go back and make sure that you read the numbers correctly from the problem. Then, double-check that you have placed the numbers in the correct spot and have labeled them properly.

Directions: Read the passage below and answer the questions that follow.

Faster, Faster, FASTER!

The setting of speed records has interested people for many years. Whether it's a car, motorcycle, bike, or even a wooden Pinewood Derby car, people have set, recorded, and broken speed records. The same is true of the competition between marathon runners and short-distance sprinters. Also, the speeds of animals such as the cheetah, the kangaroo, the ostrich, and even the exceptionally slow garden snail have been recorded.

1. The title "world's fastest woman" is traditionally held by the female athlete with the fastest run in the 100-meter sprint. In recent years, many women have burned up the track. Florence Griffith Joyner ran 100 meters in 10.49 seconds, Marion Jones in 10.65 seconds, Christine Arron in 10.73 seconds, and Merlene Ottey in 10.74 seconds. Which woman's record is closest to $10\frac{3}{4}$ seconds?

   A Florence Griffith Joyner
   B Marion Jones
   C Christine Arron
   D Merlene Ottey

2. **Part A**

   Records for the speed of cars, motorcycles, and all kinds of other modes of transportation have been set and broken many times. In 1990, a motorcycle traveled at 322.15 miles per hour. In 1995, a bicycle traveled at 166.94 miles per hour. And in 1997, a car got up to 763.055 miles per hour. Which of the three has the fastest speed?

   ____

   Part B

   Which of the speeds has a 1 in the tenths place?

   ____

3. In 1978, Kenneth Peter Warby set an official world water speed record in a powerboat by attaining 275.8 knots (equivalent to 317.6 miles per hour). Which of the following is the fractional representation of the speed in knots?

A $275\frac{1}{2}$

B $275\frac{2}{3}$

C $275\frac{3}{4}$

D $275\frac{4}{5}$

4. Cheetahs have been clocked at over 60 miles per hour for short distances. The fastest speed recorded for a kangaroo was 40 miles per hour. Ostriches can reach speeds of 37 miles per hour while running on land. The little garden snail? Well, its electrifying speed is 0.03 miles per hour. Phew, what a speed demon! If a snail were crawling at its top speed for exactly one hour, about what percent of a mile would the snail have traveled?

A 0.3%
B 3%
C 30%
D 300%

Subject Review

This chapter reviewed the place values in decimals—tenths, hundredths, and thousandths. You converted percents to decimals and decimals to percents. You converted fractions to decimals and converted decimals to fractions too.

You compared and ordered numbers, which can be a valuable skill when working with fractions, decimals, and percents. Remember what you've learned here, because the next chapter will involve fractions, decimals, and percents again.

Oh, yes, and surely you noticed that the lightning-fast speeds listed for cars, motorcycles, and bikes was kind of a surprise. But here's a more likely one for you: in 1998, a race car qualifying for the Indianapolis 500 reached a speed of 237.498 miles per hour. Wow!

Now, without further ado, the answers to the questions from page 35:

Who is the world's fastest woman?

Florence Griffith Joyner was the world's fastest woman at the time of her death in 1998. She set a women's world record in the 100-meter dash in 1988, an electric 10.49 seconds.

How many ostrich eggs would you need to make an omelette?

Not many! In fact, a single ostrich egg would produce as much food as 30 to 40 regular chicken eggs! That's one big breakfast!

How fast can a snail travel?

"Fast" isn't a good word to use. Snails are incredibly slow animals. The average garden snail can hit a top speed of 0.03 miles per hour. That's about equal to 0.013 meters per second.

CHAPTER 4

Computing with Fractions, Decimals, and Percents

What article of underwear did men and women wear in the 1800s?

What sort of person could be hungry enough to eat an airplane?

What's the recipe for Roadkill Stew and Chips?

Adding and Subtracting Fractions

Fractions are a way of indicating part of a number. Fractions with the same denominators are called **like fractions.** To add or subtract like fractions, you add or subtract their numerators. For example, $\frac{23}{49} - \frac{12}{49} = \frac{11}{49}$, because you subtract 12 from 23 and leave the denominator alone.

However, if you want to add or subtract two fractions with different denominators, you need to change the fractions by determining the **least common denominator (LCD).** Try adding $\frac{2}{7}$ and $\frac{1}{3}$. You can't make a straight calculation. However, you can determine that the LCD is 21 by multiplying the denominators 3 and 7. Change $\frac{2}{7}$ to $\frac{6}{21}$ and $\frac{1}{3}$ to $\frac{7}{21}$. Then, $\frac{6}{21} + \frac{7}{21} = \frac{13}{21}$.

The least common denominator isn't always found by multiplying the denominators. Sometimes the LCD is a lower number. Look at $\frac{1}{2} + \frac{5}{6}$. You could change each fraction to have a denominator of 12, but renaming $\frac{1}{2}$ as $\frac{3}{6}$ works just as well.

Multiplying Fractions

To multiply two fractions, you must multiply the numerators by each other and multiply the denominators by each other. $\frac{4}{5} \times \frac{3}{8} = \frac{12}{40}$, because $4 \times 3 = 12$ and $5 \times 8 = 40$. You can simplify this fraction to $\frac{3}{10}$.

Dividing Fractions

To divide one fraction by another, you multiply by the reciprocal of the divisor. The **reciprocal** is the fraction turned upside down. For example, $\frac{2}{5}$ and $\frac{5}{2}$ are reciprocals. A number multiplied by its reciprocal equals 1. The fractions $\frac{1}{4}$ and $\frac{4}{1}$ are reciprocals because $\frac{1}{4} \times \frac{4}{1} = 1$. Try dividing $\frac{1}{4}$ by $\frac{1}{3}$. First, find the reciprocal of the divisor, $\frac{1}{3}$. The reciprocal of $\frac{1}{3}$ is $\frac{3}{1}$. Now, rewrite the division question as a multiplication question using the reciprocal. $\frac{1}{4} \div \frac{1}{3} = \frac{1}{4} \times \frac{3}{1}$. The last step is to multiply the fractions, which you do by multiplying the numerators and then multiplying the denominators. $\frac{(1 \times 3)}{(4 \times 1)} = \frac{3}{4}$.

Another way to divide fractions is to remember that division is the same as repeated subtraction. Suppose you want to divide 3 feet of string into pieces that are each $\frac{1}{2}$ foot long. How many pieces of string can be cut? What you would actually be doing is the following: 3 feet $-$ $\frac{1}{2}$ foot $=$ $2\frac{1}{2}$ feet $-$ $\frac{1}{2}$ foot $=$ 2 feet $-$ $\frac{1}{2}$ foot $=$ $1\frac{1}{2}$ feet $-$ $\frac{1}{2}$ foot $=$ 1 foot $-$ $\frac{1}{2}$ foot $=$ $\frac{1}{2}$ foot $-$ $\frac{1}{2}$ foot $=$ 0. You could subtract $\frac{1}{2}$ foot 6 times, which tells you that 3 feet $\div$ $\frac{1}{2}$ foot $=$ 6.

Multiplying Decimals

Multiplying decimals is a lot like multiplying whole numbers. You just have to know how to place the decimal point in the product. For example, to calculate 4.23×6.4, you would perform the multiplication as you would for whole numbers, without putting in any decimal points. First, multiply 423×4, then 423×60, and then add these two partial products. $423 \times 4 = 1{,}692$, and $423 \times 60 = 25{,}380$. Add them! $1{,}692 + 25{,}380 = 27{,}072$. Because 4.23 has two decimal places and 6.4 has one decimal place, there are a total of three decimal places. Now, you need to count off three digits in the sum of the partial products, starting from the right, and put the decimal point to the left of the third digit. Therefore, $4.23 \times 6.4 = 27.072$.

Multiplying Percents

There are a couple of ways to find a percentage of a whole number. One is to change the percent to a fraction. Try 25% of 200. You probably know that $25\% = \frac{1}{4}$, so you can update the problem to $\frac{1}{4} \times 200 = 50$. Another way to do it is to change the percent to a decimal. 25% of 200 is the same as $0.25 \times 200 = 50.00$, or 50.

Dividing Decimals

Dividing decimals is a little bit different from multiplying decimals. To divide a decimal by another decimal, convert each decimal into a whole number. For example, to calculate $164.2 \div 0.002$, you will need to convert 0.002 to a whole number. Since the 2 is in the thousandths place, you should multiply the whole number sentence by 1,000. Once you do so, you will need to solve the new problem, $164{,}200 \div 2$. The answer you receive, 82,100, is the same as the solution to the problem $164.2 \div 0.002$. Therefore, $164.2 \div 0.002 = 82{,}100$.

Multiplication doesn't always make numbers bigger. When you multiply a positive whole number by a decimal or fraction, the product is less than the whole number, not greater than the whole number.

Directions: Read the passage below and answer the question that follows.

The History of Underwear

One article of clothing that you don't often hear people talking about is underwear. Some people call them undergarments or underclothing, but it's all the same thing. Different types of underwear have been worn under clothing for hundreds of years. As the styles change, so do the undies. Did you know that until about 1920, all underwear—men's, women's, and children's—was white? It was a symbol of a pure mind.

► In the 1800s, women wore garments called corsets to make their waists look much smaller. If a woman's waist was 42 inches and she wanted to make it 10% smaller, about how many inches would the corset decrease her waist size?

A $\frac{1}{2}$ inch
B 4 inches
C 10 inches
D 54 inches

Know It All Approach

If you read the question carefully and seek out the information needed to find the answer, you'll see that the term "about how many inches" and the numbers 42 and 10% will get you where you need to go.

The question asks you to figure the percentage of a waist measurement. There are two ways you can find the answer. One way would be to make 10% into a decimal, 0.10, and multiply by the measurement. $42 \times 0.10 = 4.2$. Because the question said "about," round 4.2 to 4 inches. Another way to find the answer is to change 10% into a fraction. $10\% = \frac{1}{10}$. Then, $42 \times \frac{1}{10} = 4\frac{1}{5}$.

There were two ways to solve the problem, so you could use the second to check your answer. If you calculated your answer using the first method, you would get 4.2. But that's not one of the answer choices! No problem! The question said "about," so you can round the answer to 4.

As always, eliminate the answer choices that are clearly too small or too big to be correct. For example, you can get rid of answer choice (A) because it's too small. Answer choice (D), 54 inches, is too large. Answer choice (B) is correct.

Directions: Read the passage below and answer the questions that follow.

Hungry Enough to Eat a Plane?

Before Michel Lotito was born in 1950 in France, bicycles were only used as a method of transportation. But Lotito turned them into food! He began to eat metal and glass when he was about 9 years old. Since his mid-teens, Lotito has eaten at least 10 bicycles. Additionally, he has found other sources of metal and glass to be super tasty. Lotito has eaten about 15 shopping carts, around 7 TV sets, and 6 chandeliers. He has even devoured a whole small aircraft. At the rate of 2 pounds of metal a day, Lotito could finish eating a television in a week or two.

1. The weights of the wheel of a bike, the handlebars, and the pedals are listed below in pounds. If Michel Lotito ate these for dinner, what would be the total weight of his dinner?

$$2 \text{ pounds} + 3\frac{1}{2} \text{ pounds} + 2\frac{3}{8} \text{ pounds}$$

A 7 pounds

B $7\frac{7}{8}$ pounds

C $8\frac{7}{8}$ pounds

D $9\frac{1}{8}$ pounds

2. If $\frac{5}{12}$ of a chandelier was made out of metal and Michel Lotito chose only to eat $\frac{1}{5}$ of the metal of the chandelier, then how much of the total chandelier would Michel Lotito have eaten?

A $\frac{1}{12}$

B $\frac{1}{5}$

C $\frac{6}{17}$

D $\frac{25}{60}$

3. The Cessna, which is a small airplane, weighs about 2,500 pounds. If Michel Lotito and a bunch of his friends ate 75% of a Cessna for lunch, how many pounds would their lunch weigh?

____

Directions: Read the passage below and answer the questions that follow.

You Just Ate WHAT?

Some shows on television have shown people eating things that are absolutely disgusting: cockroaches, grubs, and creepy crawlers that many people never even knew existed. Well, none of this is actually a new phenomenon. Long ago, people would drink gross concoctions. One ancient recipe recommended soaking 9 quarts of snails and a quart of worms in different beverages. Before drinking, the chef strained the snails and worms out of the drink. Mmm, mmm, good! A more recent addition to the gross food category is from Italy and is called "rotten cheese." It consists of—you didn't just eat, did you?—moldy, rotting curds of cheese filled with squirming maggots. Yum!

4. Did you know you can make a healthy meal that looks like barf? It's called Violet Vomit. All you need is $\frac{1}{2}$ cup corn flakes cereal, $\frac{1}{3}$ cup applesauce, $\frac{1}{3}$ cup cooked oatmeal, $\frac{1}{4}$ cup milk, 3 drops of red food coloring, and 2 drops of blue food coloring. What is the total amount of Violet Vomit, not including the food coloring, once all the ingredients have been mixed together?

   A $1\frac{1}{3}$ cups

   B $1\frac{5}{12}$ cups

   C $1\frac{1}{2}$ cups

   D $1\frac{3}{4}$ cups

5. You can make a Staring Bloodshot Eyeballs sandwich. Spread one piece of bread with lots of mayonnaise. Place slices of 2 hard-boiled eggs on the mayonnaise. Then, squeeze ketchup into lines coming out of the yolks of the eggs to make veins. If you wanted to serve 6 of your friends Staring Bloodshot Eyeballs sandwiches and each friend wanted to eat a total of 3 eggs, then how many sandwiches would you need to make?

______________________ sandwiches

6. Instead of making Staring Bloodshot Eyeballs sandwiches for your friends, you could make Roadkill Stew and Chips with help from an adult. Chop up 6 hot dogs, 2 pickles, and 3 tomatoes. Put the chopped hot dogs, pickles, and tomatoes into a microwave-safe bowl along with as much mustard and ketchup as you like. Microwave until everything is warm. Stir in a ton of grated mozzarella, and microwave again until the cheese is gooey in order to make your stew. Serve in bowls along with tortilla chips. Based on the chart below, what is the best estimate of the cost of the ingredients in Roadkill Stew and Chips?

Cost of Ingredients	
6 hot dogs	$2.89
2 pickles	$0.55
3 tomatoes	$1.10
ketchup	$1.30
mustard	$0.73
grated mozzarella	$3.74
Tortilla chips	$1.70

A $9
B $11
C $13
D $15

Subject Review

Phew! You covered a lot of information in this chapter. Don't forget to change the fractions so that they have the same the denominator when you're adding or subtracting unlike fractions. When multiplying fractions, multiply numerators together and then denominators together. To divide a fraction, multiply by its reciprocal.

Make sure you carefully count the decimal places when multiplying decimals. If there are four places in the factors, there must be four places in the product. When dividing decimals, you should convert each decimal into a whole number by multiplying by 10, 100, 1,000, or so forth, depending on how many places the smallest decimal extends.

Here are the answers you've been waiting for!

What article of underwear did men and women wear in the 1800s?

Men and women both wore an article of underwear called a corset. You may never have heard of it before. They can be made so tight that it's very difficult to breathe if you're wearing one.

What sort of person could be hungry enough to eat an airplane?

Frenchman Michel Lotito's wacky appetite has earned him a place in the history books. But they're supposed to have such good food in France!

What's the recipe for Roadkill Stew and Chips?

The ingredients for Roadkill Stew and Chips are 6 hot dogs, 2 pickles, 3 tomatoes, ketchup, mustard, grated mozzarella cheese, and tortilla chips. The actual recipe is on page 49.

CHAPTER 5

Ratios and Proportions

How many cans of soda pop does it take to build an Italian landmark?

What's a famous place to get a hot dog in Coney Island?

What roller coaster has a minor league baseball team named after it?

A **ratio** is used to compare one number to another number. Suppose that there are 25 students and 1 teacher in your class. The ratio of teachers to students is 1 to 25. Ratios can be expressed in a number of different ways. 1 to 25, 1:25, 1 out of 26, and $\frac{1}{26}$ are all ways of writing the same ratio.

A **proportion** is an equation that shows that two ratios are equivalent. An example would be $\frac{5}{10} = \frac{1}{2}$.

Directions: Read the passage below and answer the question that follows.

You Can Do It!

Ever think of building a model of a church with empty beverage cans? Well, it was done in Padua, Italy, in 1992, using more than 3,245,000 cans! The church itself, called the Basilica di Sant'Antonio di Padova, was built around 1290. The church is shaped like a cross and features massive domes and octagonal turrets.

▶ The ratio of the size of the church to the model was 4 to 1. If the model was 96 feet long, how long is the actual church?

A 24 feet
B 240 feet
C 384 feet
D 400 feet

Know It All Approach

Pay attention to the information "4 to 1" and 96 feet and the words "how long is the actual church."

Now, look at the answer choices to see if you can use Process of Elimination to get the correct answer. You can cross out answer choice (A) immediately, because the actual church will be much larger than the 96 feet of the model. The other answer choices are more reasonable, so you'll need to solve the problem.

To solve the problem, you'll want to refer to the ratio. If the model of the church is 96 feet long, and the actual church is 4 times the size of the model, then you should multiply 96 by 4. Because $96 \times 4 = 384$, you can determine that answer choice (C) is correct.

Directions: Read the passage below and answer the questions that follow.

Coney Island, U.S.A.

The amusement parks at Coney Island in Brooklyn, New York, reached their peak of popularity in the 1920s. On weekends and holidays, millions of people would head to the park to go on the innovative rides, see the bizarre sideshow, and play the games. The beach area also drew large crowds.

One of the most popular shows at Coney Island was called the Blowhole Theater. When riders would exit the Steeplechase Ride, a ride with mechanical horses, they would have to cross a brightly lit area. There, a cowboy and a dwarf would wait until a couple came by. The dwarf shocked the man with an electric stinger, and as the woman came to his aid, a powerful blast of air would blow her skirt above her waist. The people watching would scream with laughter. As couples left the embarrassing area, many would take seats to become part of the audience, laughing and enjoying the show.

1. If in one day 60,000 people visited Coney Island and 55,000 of those people took a dip in the ocean, what was the ratio of those who got wet?

   A 2 out of 3
   B 5 out of 6
   C 5 out of 60
   D 11 out of 12

2. There was an observation tower at Coney Island that took people to 300 feet above land. When it was built in 1876, it was the tallest structure in the United States. A little over 35 years later, the Woolworth Building in New York City was completed, and it is almost 800 feet tall. What's the ratio of the height of the observation tower to the height of the Woolworth Building?

____

3. The sideshow, a show that featured people with unusual looks and talents, was once a very popular attraction at Coney Island. Many of the people who were performers in the sideshow stayed there for a long time, because they felt comfortable with others who were physically different. On a summer day in 1921, if 2,100 people visited the sideshow and 100 people attending bought souvenirs, what was the ratio of people who bought souvenirs to those who did not?

   A 1 to 2
   B 1 to 10
   C 1 to 20
   D 1 to 30

4. Kelly and Jane went to Coney Island in 1926. Kelly bought 4 tickets and used 3 of them. Jane bought 8 tickets. How many tickets would Jane have to use to equal the ratio of tickets bought and used by Kelly?

   A 5
   B 6
   C 7
   D 8

5. In order to get to Coney Island, many people take the subway. A map of Coney Island uses a scale of $\frac{1}{2}$ inch = 100 meters. The map shows a distance of $2\frac{1}{2}$ inches from the subway to the ocean. Based on the scale of the map, how far is it from the subway to the ocean?

6. The same map shows that it is $1\frac{1}{2}$ inches from the ocean to the Nathan's Famous hot dog stand, one of the most popular stops at Coney Island. Based on the scale of the map, how far is it from the ocean to Nathan's?

____

7. The Cyclone roller coaster is one of Coney Island's most famous rides. There's even a minor league baseball team named after it. The map of Coney Island with a scale of $\frac{1}{2}$ inch = 100 meters shows that it is $1\frac{3}{4}$ inches from Nathan's to the Cyclone. Based on the scale of the map, how far is it from the famous hot dog stand to the famous roller coaster?

A 350 meters
B 375 meters
C 3,500 meters
D 3,750 meters

Subject Review

Chapter 5 was about ratios and proportions. Remember that a ratio is a comparison of two numbers. Proportions are equations showing that two ratios are equal.

Remember those questions from page 51? Ready for the answers? Well, here they are!

How many cans of soda pop does it take to build an Italian landmark?

A model of the Basilica di Sant'Antonio di Padova was built with more than 3,245,000 cans! That's more than 15 times the number of people who live in the city of Padua today!

What's a famous place to get a hot dog in Coney Island?

Nathan's Famous has been a hot-dog-serving institution in Coney Island since 1916. They make good French fries too!

What roller coaster has a minor league baseball team named after it?

The Cyclone is a landmark wooden-frame roller coaster in Coney Island. The Brooklyn Cyclones, whose stadium is right down the block, are a minor league baseball team operated by the New York Mets.

CHAPTER 6

Factors, Multiples, and Divisibility

How long is a goldfish's memory?

Who made the first flag of the United States of America?

How big is the flag used by the United States Navy, the Marines, and the Coast Guard?

Factors

Factors are numbers that divide equally into a whole number with a remainder of zero. The numbers 2, 4, 5, and 10 are all factors of 20. Because it is a whole number with more than two factors, 20 is called a **composite number.** The numbers 2 and 5 are **prime numbers** because they have only two factors, 1 and themselves ($2 \times 1 = 2$ and $5 \times 1 = 5$). Because they are also factors of 20, 2 and 5 are **prime factors** of 20.

Multiples

A **multiple** of a number is the product of the number and any whole number. For instance, 20 is a multiple of 4 because $4 \times 5 = 20$. Multiples are used to find the least common multiple for common denominators in subtraction problems involving fractions. Another use for multiples might be that you're a budding artist and want to have the same number of colored pencils, colored chalk, and markers on hand. Colored pencils are sold 5 to a package, colored chalk comes 15 to a package, and there are 6 markers in a package. The least common multiple for these three numbers is 30.

Divisibility

The term **divisible** means that a number can be evenly divided by another number. For example, 36 is divisible by 2. It's also divisible by 3. You might want to plant 3 rows of tomato plants in your garden. If you bought 12 plants, would there be an even number in each row? Yes, because 12 is divisible by 3.

Directions: Read the passage below and answer the question that follows.

Remember Me?

Many people have goldfish as pets. But did you know that many people believe that a goldfish's memory lasts for only 3 seconds? That means that a goldfish peering out one side of an aquarium would probably forget what it saw by the time it swam to the other side of the aquarium, or that a goldfish reading this passage would forget everything it read at the end of each sentence. Did you know that a goldfish's memory lasts for only 3 seconds? Gotcha! I bet you remembered that one.

► In a pet store, the goldfish are fed every 12 hours, and the parrots are given a treat every 44 hours. If the goldfish are fed and the parrots are given a treat at the same time, how many hours will it be until the next time these happen at the same time?

Show your work.

____

Know It All Approach

After reading the question carefully, take note of "12 hours," "44 hours," "same time," and "how many hours will it be until the next time these happen at the same time."

Now, you have everything you need. Under "Show your work," you'll want to find the multiples of 12 and 44 to see when they coincide, as follows:

12, 24, 36, 48, 60, 72, 84, 96, 108, 120, **132,** 144, . . .

44, 88, **132,** 176, . . .

It would be 132 hours until the goldfish getting fed and the parrots getting a treat coincide again.

Directions: Read the passage below and answer the questions that follow.

The First American Flag

For many years, students learned that the first flag of the United States of America was sewn by Betsy Ross at the request of George Washington. Like many other good stories, it is a legend, and one that probably will never be proved one way or the other. The story goes that Betsy Ross was asked to make a flag in 1776 that contained a design with 13 stars and 13 stripes, representing the 13 colonies. But it seems that this story was one that was passed along through the years by the descendants of Betsy Ross. There is no record that George Washington or a committee of Congress ever approached her to make the flag.

The only record that can be found regarding the flag was a bill presented to Congress in 1780 by a man named Francis Hopkinson in which he claimed to have designed the flag of the United States of America as well as the Great Seal and was due payment. Congress determined that he didn't act alone in the design and that any payment should be shared. They argued about paying him for a long time, but they never denied that he had designed the flag.

1. Of the 13 stripes on the American flag, 7 of the stripes are red and 6 of the stripes are white. If the standard size of the flag used by the Navy, Marines, and Coast Guard is 52 inches wide, what is the width of each stripe?

   A 3 inches
   B 3.5 inches
   C 4 inches
   D 4.5 inches

2. As more states became part of the United States, more stars were added to the flag. In 1846, there were 28 stars. The factors of 28 will indicate the different ways the stars could have been arranged. Find the factors of 28.

____

3. From 1912 to 1959, there were 48 stars on the flag. Which of the following was a possible arrangement of the 48 stars?

A 2 rows of 22 stars
B 3 rows of 18 stars
C 4 rows of 24 stars
D 6 rows of 8 stars

4. Which of the following are the prime factors of 48?

A 2 and 3
B 4 and 12
C 2, 4, and 12
D 2, 12, and 24

5. Since 1960, there have been 50 stars on our flag. What are all of the factors of 50?

____

6. For a parade on the Fourth of July, a group of marchers will carry the flag of the United States. The members of the group want to arrange themselves in rows of 4, 6, or 8. Every time they arrange themselves in rows of 4, 6, or 8, there are 3 members left over. Which of the following could be the total number of marchers in the group?

A 19
B 39
C 51
D 107

Subject Review

In this chapter, you learned about factors, multiples, and divisibility. Remember that factors are numbers that divide evenly into a whole number, and prime factors are factors that are only divisible by 1 and themselves. A multiple is the product of a number and a whole number. Divisibility means that a number can be evenly divided by another number.

Now, here are the answers from the questions at the beginning of the chapter.

How long is a goldfish's memory?

The goldfish has an extremely short memory, only 3 seconds. By the time you finish reading this sentence, a goldfish will forget that you started it!

Who made the first flag of the United States of America?

Some people believe that the first flag was made by Betsy Ross, and others believe that it was made by Francis Hopkinson. This is one of those facts from history that will probably never be answered to everyone's satisfaction.

How big is the flag used by the United States Navy, the Marines, and the Coast Guard?

The flag commonly displayed by the Navy, Marines, and Coast Guard is 52 inches wide and 66 inches long.

CHAPTER 7

Exponents and Square Roots

How long have there been lotteries?

How did something called a coelacanth shock scientists in 1938?

What insect has been on earth for more than 300 million years and can live for up to several weeks after its head has been cut off?

Exponents

In mathematics, you can often use short forms to write ideas in a more compact way. There is a shorter way to show multiplication when all of the factors are the same number. For example, there is a shorter way of writing $2 \times 2 \times 2 \times 2$. You can use an exponent to write the above expression. An **exponent** is a way of showing how many times a factor is multiplied by itself. In $2 \times 2 \times 2 \times 2$, the factor, 2, is multiplied by itself 4 times. You could write this as 2^4. In this example, 2 is the base. The **base** is the factor to be multiplied by itself. The exponent is 4, because the base is multiplied 4 times. The entire expression 2^4 is called a **power.** The expression 2^4 is written in **exponential form.**

power { 2^4 ← **exponent**
↑
base

Scientific Notation

Scientific notation is a way of writing very large or very small numbers. In scientific notation, a number is written as the product of a factor and a power of 10. For example, the number 90,000,000 is written as 9×10^7 in scientific notation. The reason that the exponent for the base 10 is 7 is because there are 7 digits after the 9 in 90,000,000. The number 43,500,000,000 is written as 4.35×10^{10} in scientific notation. Count the number of places the decimal has moved to the right from between the 4 and 3 to the last zero. It's 10 places, the same as the exponent.

Square Roots

When you multiply a number by itself, the result is the square of the number, or a **perfect square.** For example, 25, which is 5×5 or 5^2, is a perfect square. You can work backwards from a perfect square to find its square root. A **square root** is the number representing both of the factors of a perfect square. Look at the picture below.

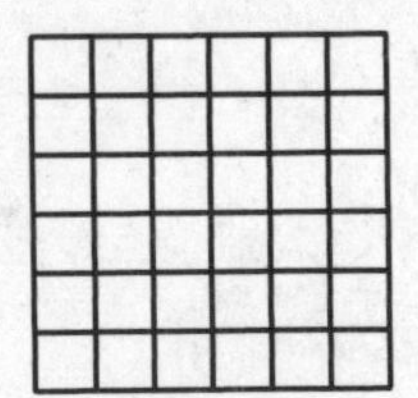

This picture shows a large square made up of 36 smaller squares. Each side of the square is made up of 6 smaller squares, so each side is 6 units. You can write the following multiplication sentence for the picture: $6 \times 6 = 36$. The square root of 36 is 6, because both factors of 36 are 6.

To show a square root, you use a **radical sign** ($\sqrt{}$). Some square root questions may use the radical sign. Other questions will simply ask for the square root of a number.

Directions: Read the passage below and answer the questions that follow.

Dream Big

You may think that lotteries are a modern invention. Not so. They've been around for hundreds of years. The first lotteries in Europe were organized in the fifteenth century to raise money to strengthen the military and to help the poor. Queen Elizabeth I used a lottery in 1566 to raise funds for restoring harbors. In the United States, in 1776, lotteries were used to raise money for the American Revolution. By the early 1830s, lotteries had become extremely popular and were used in 8 states, which is a third of the number of states that existed at that time.

▶ In December 2002, a West Virginia man named Andrew Jackson Whittaker, Jr., won $314.9 million, or $314,900,000, in a single lottery jackpot. How would you write this number in scientific notation?

____

Know It All Approach

Read the question carefully and try to find out what you'll need to solve this problem. The number 314,900,000 and the term "scientific notation" are the key pieces of information.

In this question, you're not really calculating the answer. Instead, you're counting the number of places from behind the last zero to between the 3 and 1. That's 8 places.

Now, put the number in scientific form: 3.149×10^8.

Check your answer for this problem by making sure that you wrote the number correctly and counted the exact number of places. Finally, write the answer on the line provided.

Directions: Read the passage below and answer the questions that follow.

A Living Fossil

Did you ever hear of a living fossil? Scientists depend on fossils to tell them the story of something that became extinct long ago. Living fossils, however, are animals that never became extinct and have barely changed over time. A coelacanth (SEE-luh-kanth) is a living fossil. The coelacanth was a prehistoric fish that was known to grow to sizes of several feet in length and more than 100 pounds. Until 1938, the only reason scientists knew about the coelacanth was because of fossils, the most recent of which was 136 million years old.

Imagine, then, how amazed the scientist Marjorie Courtenay-Latimer was in 1938 when some local fishermen brought her the strange-looking fish they had just caught on a fishing trip! Although Courtenay-Latimer didn't know what it was at first, the fish was later confirmed as a coelacanth. It was an amazing discovery for all in the scientific community.

1. A huge extinction occurred on the earth 65 million years ago. Millions of species of animals and plants, including the dinosaurs, vanished. What is 65 million (65,000,000) in scientific notation?

   A 6.5×10^5
   B 6.5×10^6
   C 6.5×10^7
   D 6.5×10^8

2. The horseshoe crab is an example of a prehistoric animal that has survived into the modern era. In fact, the horseshoe crab isn't a crab at all. It is more closely related to spiders and scorpions. Female horseshoe crabs lay their eggs in a series of shallow sand pits during the spring high tides. To find the approximate number of eggs they lay in each pit, solve the following:

   $5^3 \times 2 =$ ______________ eggs

3. The tuatara is a reptile that is also called a beakhead because it has a hooked jaw like the beak of a bird. When the tuatara is cold, it may take only one breath over the course of an entire hour. Most people would turn blue: How about you? Their heartbeat also slows down. Find the average number of heartbeats per hour of a tuatara by finding the perfect square between 42 and 55.

____

4. The cockroach has been around for more than 300 million years. It is able to survive many poisons, can live even after its head has been chopped off, and isn't bothered by high levels of radiation. It will eat anything! The inside of its stomach has teeth that grind up whatever it eats. Digested food is excreted through 60 different tubes. Oh, yuck! If the 60 tubes were unfolded, they would cover 1.32×10^5 millimeters. Which of the following is 1.32×10^5 in standard form?

A 13,200
B 132,000
C 1,320,000
D 13,200,000

5. A darkly armored beast floats down the Nile River with only its eyes above the surface. It seems so slow and lazy, not paying any attention to the small animals on the riverbank. Then, suddenly, SNAP! The crocodile has leaped out of the water and captured a heedless animal in its strong jaws. Dinner is served. The length of this behemoth is the square root of 324 feet. How long is the crocodile?

A 12 feet
B 14 feet
C 16 feet
D 18 feet

Subject Review

This chapter explained exponents, scientific notation, and square roots. Exponents tell you how many times to multiply the base number by itself. Scientific notation is a much shorter way to express very large numbers. Square roots are the reverse of squaring a number, or multiplying it by itself.

Want the answers to the questions from the beginning of the chapter? Here they are!

How long have there been lotteries?

Historians confirm that lotteries have been around since at least the 1400s. Even if you played the lottery every week since the year 1400, you still probably wouldn't have won very much.

How did something called a coelacanth shock scientists in 1938?

The coelacanth was a prehistoric fish that scientists thought was extinct until some fisherman caught one near South Africa in 1938. The species had been swimming around all along, for more than 286 million years!

What insect has been on earth for more than 300 million years and can live for up to several weeks after its head has been cut off?

The icky cockroach. Yuck!

CHAPTER 8
Properties of Numbers

What does the most famous doll in history have on her tummy?

Who ate $50\frac{1}{2}$ hot dogs in 12 minutes?

In mathematics, there are many rules that always hold true regardless of the numbers being calculated. These rules are referred to as the **properties of numbers.**

Associative Property

The **associative property** tells you that no matter how you group the numbers when you add or multiply, you will get the same answer.

$(5 + 4) + 6 = 5 + (4 + 6)$

$\left(\frac{1}{2} + \frac{2}{3}\right) + \frac{5}{6} = \frac{1}{2} + \left(\frac{2}{3} + \frac{5}{6}\right)$

$(1.2 + 2.6) + 4.2 = 1.2 + (2.6 + 4.2)$

$(5 \times 4)6 = 5(4 \times 6)$

$\left(\frac{1}{2} \times \frac{2}{3}\right)\frac{5}{6} = \frac{1}{2}\left(\frac{2}{3} \times \frac{5}{6}\right)$

$(1.2 \times 2.6) \times 4.2 = 1.2 \times (2.6 \times 4.2)$

Commutative Property

The **commutative property** tells you that you can switch the order of numbers in an addition or multiplication problem and still get the same result.

$9 + 7 = 7 + 9$

$\frac{4}{7} + \frac{3}{5} = \frac{3}{5} + \frac{4}{7}$

$7.1 + 3.6 = 3.6 + 7.1$

$(9)(7) = (7)(9)$

$\left(\frac{4}{7}\right)\left(\frac{3}{5}\right) = \left(\frac{3}{5}\right)\left(\frac{4}{7}\right)$

$7.1 \times 3.6 = 3.6 \times 7.1$

Distributive Property

The **distributive property** is used when you have expressions that contain both addition *and* multiplication. If you need to multiply a single number by a group of numbers inside parentheses, you can distribute the multiplying number to each of the numbers inside the parentheses.

$3(2 + 8) = 3(2) + 3(8)$

$\frac{1}{4}\left(\frac{4}{5} + \frac{3}{7}\right) = \frac{1}{4}\left(\frac{4}{5}\right) + \frac{1}{4}\left(\frac{3}{7}\right)$

$3.8(1.3 + 4.5) = 3.8(1.3) + 3.8(4.5)$

$(2 + 8)3 = (2)3 + (8)3$

$\left(\frac{1}{4} + \frac{4}{5}\right)\frac{3}{7} = \left(\frac{1}{4}\right)\frac{3}{7} + \left(\frac{4}{5}\right)\frac{3}{7}$

$(3.8 + 1.3)4.5 = (3.8)4.5 + (1.3)4.5$

Directions: Read the passage below and answer the question that follows.

Tattoo You!

The art of tattooing has been around for a very long time. In fact, a mummy found in the mountains between Austria and Italy who died around 3300 B.C. has markings on his back and ankles that scientists think might be tattoos. Some ancient Egyptian mummies also have tattoos. The first electric tattooing machine was invented in 1891. Before then, elaborate tattooing was a symbol of rank and wealth because it was too expensive for most people to pay the artist. Today, celebrities, musicians, sports figures, and even Barbie have tattoos.

▶ Which expression uses the distributive property to find the total area of the tattoo design below? (You find the area of a square or rectangle by multiplying its length by its height.)

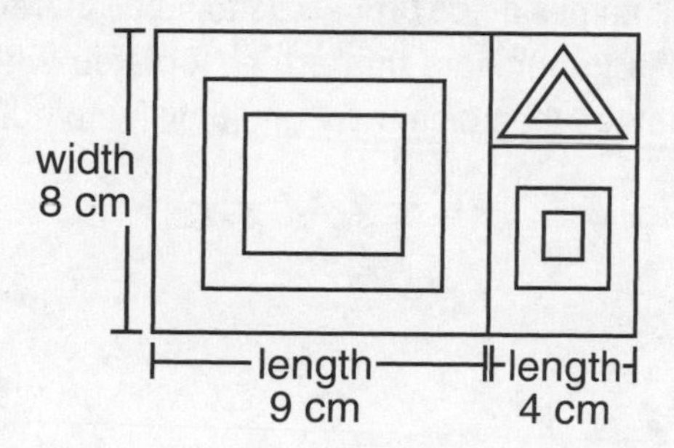

A $8 \times 9 \times 8 \times 4$
B $8(9) + 8(4)$
C $8 + (9 \times 4)$
D $8(9 \times 4)$

Want to learn more about the mummy found in the mountains? Go to page 100.

Know It All Approach

After reading the question carefully, take note of the term "distributive property" and the phrase "area of the tattoo design."

The formula for the area of a rectangle is length times height. Don't worry if you don't remember this. Chaper 12 reviews area. In this case, you have the width of the rectangle, which is 8 inches. The length is 9 inches + 4 inches. So, to find the area, you would have $A = 8 \times (9 + 4)$. This answer uses the distributive property because $8 \times (9 + 4)$ is the same as $8(9) + 8(4)$.

Answer choice (A) is incorrect because all of the dimensions are multiplied, and the two lengths are not added. Eliminate it. Answer choices (C) and (D) do not use the correct formula for area. You can toss them out as well. Answer choice (B), however, uses the distributive property and is identical to the answer you calculated.

Think about the names of the properties to help you remember them. The associative property applies to numbers that are associated with each other. The distributive property distributes the number you multiply by to other numbers.

Directions: Read the passage below and answer the questions that follow.

Buns and All

Sometimes professional athletes are accused of cheating in order to win competitions. But have you ever heard of anyone accused of cheating in an eating competition? In 2002, Takeru Kobayashi ate $50\frac{1}{2}$ hot dogs in 12 minutes to win a hot-dog-eating contest at Nathan's Famous in Coney Island, New York. The year before, he finished 50 hot dogs, which at the time was about twice as many hot dogs as any competitor had ever eaten in the contest.

Some people believed he had cheated. How does someone cheat during a hot-dog-eating contest? Kobayashi was accused of regurgitating, meaning that some thought he had spit up some of the hot dogs he supposedly ate before finishing the contest. Kobayashi started out the competition weighing about 113 pounds and finished it weighing about 120 pounds, so he must've been painfully stuffed. But video footage shows that the liquefied food that shot out of stuffed Kobayashi's nose didn't come out until after the contest was completed.

1. In 2002, Kobayashi ate 50.5 hot dogs. If 1 hot dog weighs 0.13 pounds, then which of the following shows the commutative property for finding the weight in pounds of the hot dogs Kobayashi ate?

   A $50.5 + 0.13 = 0.13 + 50.5$
   B $(50.5 - 0.13)0.13 = (50.5)0.13 - (50.5)0.13$
   C $(50.5)(0.13) = (0.13)(50.5)$
   D $(50.5 + 0.13)(50.5 + 0.13) = (50.5 + 0.13)^2$

2. In 2002, Kobayashi ate 50.5 hot dogs, another competitor ate 26 hot dogs, and another competitor ate 20 hot dogs. Which of the following equations could be used to find the total number of hot dogs eaten, *H*, by Kobayashi and those competitors?

   A $3(50.5 + 26 + 20) = H$
   B $50.5 + (26 + 20) = H$
   C $3(50.5) + (26 \times 20) = H$
   D $50.5 + (26 \times 20) = H$

3. You and your friend are in the mood to eat hot dogs for lunch. You buy a hot dog for $1.20, a juice for $0.75, and an apple for $0.35. Your friend buys a hot dog for $1.20, a juice for $0.90, and a banana for $0.25. Use the associative property to write an expression for the total cost of what you and your friend bought for lunch.

   ____

Subject Review

In this chapter, you have used the associative, commutative, and distributive properties. Each of the properties can be used for integers, decimals, and fractions. The associative property means that you will get the same answer no matter how you group the numbers in addition or multiplication. The commutative property lets you change the order of the numbers in addition or multiplication and still get the same answer. The distributive property takes the number you multiply by and distributes it to the numbers grouped in parentheses.

Take a look at the answers to the questions from the beginning of the chapter.

What does the most famous doll in history have on her tummy?

A Barbie doll issued in 1999 featured a butterfly tattoo. However, due to several complaints, the doll was taken off the shelf. It is now a collector's item.

Who ate $50\frac{1}{2}$ hot dogs in 12 minutes?

The 24-year-old Japanese citizen Takeru Kobayashi accomplished this amazing feat at the annual Fourth of July hot-dog-eating contest in Coney Island, New York.

Directions: Read the passage below and answer the questions that follow.

Curly

Do you know what famous music star has the nickname Curly? It's Justin Timberlake, who with the band 'N Sync became a sensation in the 1990s. Timberlake was born in 1981 and showed his interest in music at an early age by singing in harmony by the time he was $2\frac{1}{2}$ years old. He became a choir member at church and competed in various talent shows to gain experience. At the age of 12, he was chosen to be on the Mickey Mouse Club show. There he met one of his fellow 'N Sync members, JC (Joshua Chasez), as well as Christina Aguilera and Britney Spears. After the show was cancelled, 'N Sync was formed and began performing and recording.

1. In March 2000, 'N Sync released its album *No Strings Attached*. It sold an amazing 2.4 million copies in the first week. What is 2.4 million (2,400,000) written in scientific notation?

   A 2.4×10^4
   B 2.4×10^5
   C 2.4×10^6
   D 2.4×10^7

2. It is estimated that *No Strings Attached* sold 1.1 million copies in the first day after its release. If it had continued to sell at that pace, how many copies would it have sold after 30 days?

   ____

3. 'N Sync's older CDs can be purchased used. The list below shows some of the used prices of each CD. If all of the CDs were purchased, what is the best estimate of their total cost?

'N Sync Used Albums

'N Sync	$1.95
No Strings Attached	$5.95
Celebrity	$4.35
Home Christmas	$3.05
Winter Album	$2.89

A $11.00
B $14.00
C $16.00
D $18.00

4. Which number best represents the dot on the number line, which is Justin Timberlake's height?

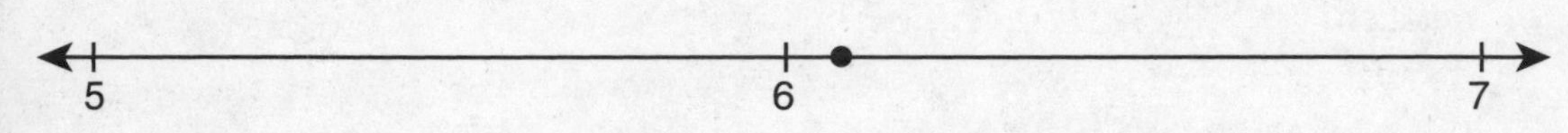

A $5\frac{7}{8}$
B $6\frac{1}{12}$
C $6\frac{1}{3}$
D $6\frac{1}{2}$

5. Justin Timberlake released a solo album called *Justified.* If it sold 20,000 copies on the day it was released and 'N Sync's album *Celebrity* sold 30,000 copies on the day it was released, what is the ratio of first-day sales for *Justified* to those for *Celebrity*?

____

6. 6 copies of *Justified* and 3 copies of *No Strings Attached* were bought as a prize for a skateboard contest for $16.99 each. Write an expression that uses the distributive property to find the total cost of the albums.

____

Directions: Read the passage below and answer the questions that follow.

The Story of Blackbeard

Does the name Edward Teach send shivers down your spine? How about the name Blackbeard? Edward Teach was the notorious pirate called Blackbeard for his thick, bushy black beard. There are stories that Blackbeard would twist his beard into tangled loops and insert burning matches, which would wreath his head in smoke. He always carried an ample supply of pistols, swords, and daggers.

Blackbeard attacked many ships, stealing their cargo and the money and jewelry of the people on board. He would prey on merchant ships as they left safe harbors. He and his ships were considered a menace, and soon people decided to do something about his evil ways. He was aboard his ship when it was attacked by a British Navy ship near Ocracoke, an island off the coast of North Carolina, in 1718. During the fighting, he was killed, and his head was put on a pole at the front of the British vessel.

7. Julian read 12 books about his favorite pirate, Blackbeard. Which of the following are factors of 12?

   A 2, 3, 4, 6
   B 2, 3, 4, 8
   C 2, 3, 4, 5, 6
   D 2, 3, 4, 6, 8

8. Pirates sailing the Caribbean in the early 1700s numbered between 44^2 and 45^2. Which of the following could be the number of pirates? (Use your calculator if necessary.)

   A 1,800
   B 1,920
   C 2,000
   D 2,030

9. Blackbeard's flagship was called the *Queen Anne's Revenge.* On its decks rode 40 guns, and 300 men were aboard. What was the ratio of men to guns?

   ____

10. At one time, Blackbeard took hostages and demanded ransoms. If he took 18 hostages and $\frac{2}{3}$ of them were men, how many men were there?

    ____

11. A pirate with the reputation of Blackbeard would have taken a lot of stuff from the ships he raided. Some legends associated with more successful pirates say that they buried some of their treasure and no one ever found it. A map leads to one such treasure. The scale of the map is $\frac{1}{2}$ inch $=$ 100 feet. If the distance from Quicksand Point to Skull Cliff on the map is 2 inches, what is the actual distance?

    A 20 feet
    B 200 feet
    C 300 feet
    D 400 feet

CHAPTER 9

Variables and Linear Equations

Why did archaeologists in Australia dig for several months straight?

Who is the real Indiana Jones?

How fast is a velociraptor?

Variables

A **variable** is a symbol (usually a letter) that represents a number. In the equation $8n + 7 = 31$, the variable is n. It represents a number. To find the value of the number, you have to solve the equation for n.

Variables are used in formulas too. For example, the formula to determine the area of a rectangle is $A = lw$. A is the area, and l and w stand for the length and the width of the rectangle. You'll learn more about area in chapter 12. If you know the area is 56 and the length is 8, you can change the formula to $56 = 8w$. Solving for the variable w will determine the width.

There may be more than one variable in an expression or equation. The values of the two variables in an expression or equation are dependent on each other. For example, in the equation $y = 2x$, the value of y is dependent on the value of x, and the value of x is dependent on the value of y.

Equations

An **equation** is a number sentence that sets two values that are equal to each other on either side of an equal sign. For example, imagine that you save $10 each week from a job mowing your neighbor's lawn. You want to know how long it will take you to save enough money to buy a cool jacket that costs $160. You'd use this equation: $10w = 160$. Solve for w to find out how many weeks you would need to work.

Divide both sides of the equation by 10, as follows: $\frac{10w}{10} = \frac{160}{10}$. Then, you can see that $w = 16$. It would take you 16 weeks to save the money. Now, say you got $40 for your birthday. How long would it take you to save $160 then? First, add 40 to the first part of the equation. $40 + 10w = 160$. Then, subtract 40 from both sides. $10w = 120$. Divide both sides by 10 to find that $w = 12$. With the gift, it would take you 12 weeks to save $160 for the jacket.

Graphing Linear Equations

Some equations have two variables that form a straight line when graphed on a coordinate grid. Such equations are called **linear equations.** A linear equation can also be represented as a word problem. For example, at the Gobble Gas pumps, a gallon of gas is pumped every 2 minutes. (Hope you're not in a hurry.) To find the amount of time it would take to fill a car's tank, you could use the equation $y = 2x$, where the y stands for gallons and the x stands for time.

Use a function table or substitute values. Start with the one you know: If $x = 1$, then $y = 2(1)$, or 2. If $x = 2$, then $y = 2(2)$, or 4. If $x = 3$, then $y = 2(3)$, or 6. If $x = 4$, then $y = 2(4)$, or 8. The time is always twice the number of gallons. See what this information looks like when it's graphed on the coordinate grid below.

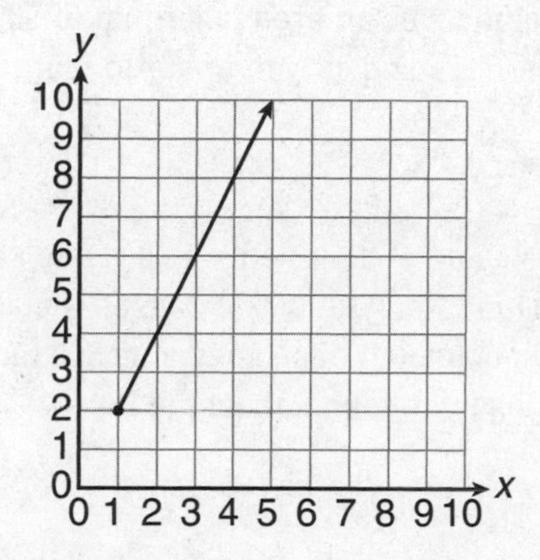

Remember that the first number in an ordered pair is the x-coordinate and the second number is the y-coordinate. That means the first number tells you how far to the left or right a point is. The second number tells you how far up or down a point is. Keep this in mind when you graph linear equations.

Directions: Read the passage below and answer the question that follows.

Do You Dig Digging?

Have you ever dug a hole before? Maybe you've dug one at the beach or in your backyard. While there have been holes dug by dogs to bury bones, by construction workers to build houses, and by miners to find valuable ore and minerals, some people dig holes around the world to find the past! In particular, one of the largest archaeological digs ever in Australia has uncovered how some people lived in the early 1800s. Archaeologists, or people who study fossils and remains of human and animal life, dug for months in order to uncover an area where up to 500 people may have lived and worked. The archaeologists are trying to learn how people in Australia lived in the past.

▶ Over time, holes usually lose some of their depth as dirt and soil fills them in. You can use an equation to determine the fill-in rate. A landslide caused a giant hole to fill up 15 feet. For the next 57 years, it filled the same amount each year, until 300 feet of dirt was in the hole. Which equation below represents the information about the hole?

A $15 - 57x = 300$
B $15 + (57 \div x) = 300$
C $(15 - x) \div 57 = 300$
D $15 + 57x = 300$

Know It All Approach

If you read the question carefully, you should notice the information you need to answer the question. The term "fill up 15 feet" and the numbers 57 and 300 are necessary information, but don't forget the phrase "the same amount each year." That is the key to this problem.

To solve this problem, you don't have to calculate an answer. You just have to choose the correct equation from the given choices. The hole filled 15 feet and then filled x amount for 57 years. So far, you have 15 and $57x$. Because the hole gets filled, you need to add the values $15 + 57x$. The amount it filled for 57 years is the variable, and multiplying it by 57, gives you $57x$. The hole filled a total of 300 feet. Your equation is $15 + 57x = 300$.

Read the problem you were given again as you follow along with the equation. Make sure that your equation will solve the wording of the problem. Does it make sense? If so, you're in great shape.

Because your answer is one of the answer choices listed, you're in pretty good shape. You should still read through all the answer choices, just to be sure that yours is the right one. Answer choices (A) and (C) use subtraction, so they're not right. You can cross them off. Answer choice (B) uses division for finding the amount that went into the hole each year. That's not right either. Looks like answer choice (D) is the correct choice.

Directions: Read the passage below and answer the questions that follow.

The Real Indiana Jones?

Did you know that Indiana Jones was probably based on a real scientist? His name was Roy Chapman Andrews. When Andrews was a young man, his greatest dream was to work at the American Museum of Natural History in New York, because he had a great love for natural history. After graduating from college, Andrews went to New York to ask the director of the museum for a job. There were no jobs available, but Andrews insisted that he would work in the museum in any capacity—even cleaning floors. The director hired him and assigned him to sweep the floors in the taxidermy department. That's where dead animals are preserved and stuffed for display.

Over the years, Andrews worked hard, and he eventually became one of the world's most famous explorers, traveling to dangerous and faraway places, just like the movie character Indiana Jones. Among his many trips, Andrews traveled to the Gobi Desert in Mongolia, where he discovered miles and miles of dinosaur bones while searching for the remains of prehistoric humans. He had to deal with burning and freezing temperatures and survive attacks by bandits. Almost 30 years after he started working at the American Museum of Natural History, Roy Chapman Andrews became the museum's director. He was one of the world's great scientists and adventurers.

1. One of the most important discoveries from Roy Chapman Andrews's expeditions was that dinosaurs were hatched from eggs. During the expedition of 1923, his team collected several valuable eggs. Solve the equation below to find how many eggs were collected.

$$3x - 16 = 59$$

____________ eggs

2. A significant find on one of Andrews's expeditions was the fossil of a velociraptor dinosaur. This carnivore's razor claws, sharp teeth, and sheer viciousness became well known after the movie *Jurassic Park*. To find how fast it was able to run, use the formula for velocity, $V = \frac{d}{t}$, where d stands for distance and t stands for time. What is a velociraptor's speed if it goes 171 meters in 9.5 seconds?

   A 22 meters per second
   B 16 meters per second
   C 18 meters per second
   D 13 meters per second

3. One cold night during an expedition in 1925, many pit vipers, which are extremely venomous snakes, came into the tents for warmth. The explorers defended themselves and killed many of the snakes. Rewrite the sentence below as an equation, and then solve it. The solution indicates the number of snakes that were killed by the explorers.

   The quotient of 360 and 8 equals a number minus 2.

   ____________________ = __________

4. Sometimes camels were used on expeditions. Camels can go a long time without water, but when they are thirsty, they can drink 25 gallons of water in 10 minutes. The equation $y = 2.5x + 5$ shows how many gallons of water would be needed per minute for 5 very thirsty camels, using y for gallons and x for minutes. Which ordered pair below satisfies the equation?

   A (5, 17)
   B (6, 20)
   C (4, 14)
   D (3, 12)

5. The expeditions carried a variety of scientists, technicians, assistants, and cooks, who may have been in four different vehicles. Trucks 1 and 2 together carried 28 people. Trucks 3 and 4 together carried 25 people. Trucks 1 and 4 together carried 23 people, and Truck 2 carried 15 people. How many people were in Truck 3?

   A 22
   B 17
   C 24
   D 15

6. Andrew's team used crates to transport fossils back to the museum. If the perimeter of the lid of one crate is 46 inches and the length is 12 inches, what is the width of the lid? Use the formula for perimeter, $P = 2l + 2w$.

   A 11 inches
   B 9 inches
   C 15 inches
   D 12 inches

Subject Review

In this chapter, you learned about variables, such as *n,* that represent an unknown number. You also learned how to solve equations and how to graph linear equations on a coordinate grid.

Remember that variables are usually letters that take the place of a number in an expression or an equation. In an equation with a single variable, you solve the equation for the variable by isolating it on one side of the equal sign.

When you graph linear equations, you use a function table or substitute for one variable to find the value of the other. Then, you use the numbers in the function table to graph points on the coordinate plane.

Want the answers to the questions from page 81? Well, here they are!

Why did archaeologists in Australia dig for several months straight?

The archaeologists were looking for clues into the history of people who lived in Australia in the 1800s.

Who is the real Indiana Jones?

Roy Chapman Andrews was likely the inspiration for the Indiana Jones movies. And just like Indy, he hasn't had the best luck with snakes.

How fast is a velociraptor?

The fierce velociraptor dinosaur could run at speeds of 18 meters per second. That is about 40 miles per hour. You wouldn't even be able to outrun one of those guys on your bicycle!

CHAPTER 10

Patterns and Functions

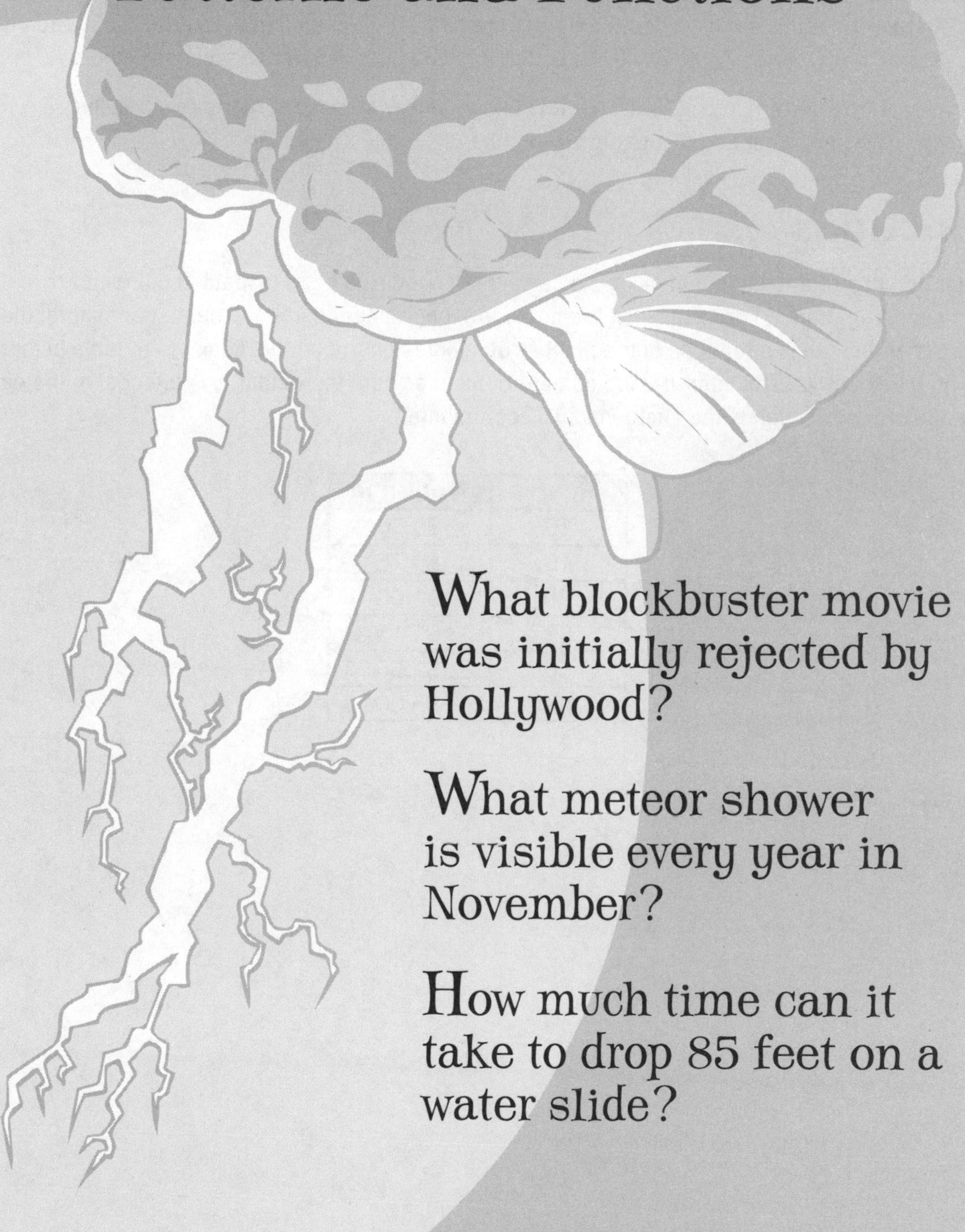

What blockbuster movie was initially rejected by Hollywood?

What meteor shower is visible every year in November?

How much time can it take to drop 85 feet on a water slide?

Patterns

A number **pattern** is a series of numbers that are related in a certain way. In the series {3, 6, 9, 12}, the pattern is that each number is 3 more than the one before it. Another example is {11, 12, 14, 17, 21}. The pattern in this series is that the difference between the numbers increases by 1 each time. The pattern in the series {5, 10, 20, 40, 80} is that each number is twice the number before it.

Can you see the pattern in {1, 1, 2, 3, 5, 8, 13}? Each number is the sum of the two previous numbers. The next number in the pattern would be 21.

Functions

A **function** is a relationship that follows a certain rule. The rule pairs up each input with a corresponding output. For example, if the cost for a long-distance call is 5 cents per minute, then the cost of the call is a function of the number of minutes on the phone. Look at the table below to see how this works. The input *n* is the number of minutes, and the output *m* is the cost of the call. The rule of the function is the charge of $0.05 per minute.

Input *n*	Output *m*
10	$0.50
15	$0.75
20	$1.00
23	$1.15
50	$2.50
100	$5.00

Directions: Read the passage below and answer the question that follows.

Nobody Wanted to See Star Wars?

Anakin Skywalker, Obi-Wan Kenobi, Yoda, R2D2. These names may be familiar to you if you're a *Star Wars* fan. When director George Lucas tried to sell his unusual movie to the major Hollywood studios, they refused to make the picture because they thought that people wouldn't want to see it. They thought the idea was silly! Finally, 20th Century Fox agreed to make the first *Star Wars* movie, and history was made. The first *Star Wars* movie was released in 1977, and the two others in the original trilogy followed in 1980 and 1983. Those three movies are still among the top money-making movies of all time.

► On a Saturday, Tariq watched *The Empire Strikes Back.* 2 days later he watched it again. 4 days later, he watched it a third time. Tariq decided to watch it again 6 days after that. If the pattern continues, how many days will it be before Tariq watches *The Empire Strikes Back* again?

A 2 days
B 4 days
C 8 days
D 11 days

Know It All Approach

After reading the question carefully, you may see that 2, 4, 6, and"pattern continues" are the items you need to notice. Now, you have to look for the pattern to answer the question. This one may be a little tricky. You need to find the pattern from the first three numbers.

Tariq watched the movie for the second time 2 days after he saw it for the first time. Tariq watched the movie for the third time 4 days after he saw it for the second time. Tariq watched the movie for the fourth time 6 days after he saw it for the third time. Each time Tariq watches the movie it is 2 days longer than the previous amount of time between viewings. Therefore, if the pattern continues, you can calculate that Tariq will wait 8 days before watching the movie a fifth time.

You can double-check your answer. Look at the figures again to make sure you've found the correct pattern. You may want to write the pattern out as a series of addition problems, such as $2 + 2 = 4$, $4 + 2 = 6$, and $6 + 2 = 8$.

Now, read all of the answer choices to determine whether your answer is there. Your answer, 8, is listed as choice (C).

You can use Process of Elimination to rule out the wrong answer choices. Choice (A) seems incorrect because it doesn't follow the pattern of 2, 4, 6, . . ., but 2 days is the amount of time added to each period between Tariq's viewings. Therefore, before you rule it out, make sure that you understand the question correctly. The question asks "how many days will it be before Tariq watches *The Empire Strikes Back* again?" Once you have reread the question and understood it, you can rule out answer choice (A). You can rule out answer choice (B) because 4 days is 2 days less than the time between Tariq's previous viewings. The pattern is built on the idea that Tariq increases the amount of time between viewings. You can rule out answer choice (D) because it is an odd number. If the pattern always involves adding 2 to an even number, there will never be a period of 11 days between viewings. Answer choice (C) is correct.

Directions: Read the passage below and answer the questions that follow.

Trivial Trivia

The world is filled with interesting trivia. There are trivia board games, trivia Web sites, and trivia contests. Some people are very good at remembering facts that seem to be hardly worth the effort. A lot of people are amazed by the minutiae, or trivial details, that they have retained. Even people who think they don't know any trivia are able to startle themselves and their friends at unexpected times. The following set of questions belongs in a trivia game. Can you find and remember the answers?

1. During the Leonid meteor shower each year in mid to late November, it is possible to see 1,000 meteors in a single hour. What a thrill! In the normal evening sky, the meteor rates are quite a bit lower. The series of numbers below shows the number of meteors per hour for one shower, starting with 63 at 11:00 P.M. Based on the pattern, what is most likely the number of meteors that fell during the missing hour?

   63, 66, 69, ________________, 75, 78

   A 3
   B 72
   C 6
   D 81

2. One type of insect lives underground for most of its life. Eventually, it comes to the surface, climbs a tree, sheds its skin, and lives for only a few weeks. This insect is known as the periodical cicada. This kind of cicada emerges at regular intervals throughout the years. Given the dates below, what is the number of years between each emergence of the periodical cicada?

   1965, 1982, 1999, 2016

   ________________ years

3. Bamboo is an exceptionally fast-growing plant. It can reach its full height in 30 to 90 days. That is good news for the hungry panda bears that like to eat it. The numbers below show the approximate height, in inches, of a young bamboo plant. Each number represents its height on the following day. What is the missing number in the series?

46, 69, ____, 115, 138

____

Directions: Read the passage below and answer the questions that follow.

Water Fun!

A water park in Ohio called the Beach features an 85-foot water slide called the Banzai. If you ride it, you will scream toward the pool in a mere 6.5 seconds. Water parks are extremely popular throughout the world. Water slides with vortexes, wave pools that make waves 6 feet high, and leisurely tube rides are all part of the fun. Caribbean Bay, a modern water park in Seoul, South Korea, broke a world record for attendance to water parks. In its first 2 years, more than 1.5 million people visited it for all of those attractions and more!

4. Imagine a zippy water slide that increases the amount of water that flows as you move farther and farther down the slide. What a ride! The table below indicates the rate at which the water flows. If the pattern in the table continues, how many liters of water will be released after 10 seconds?

Water Release

Seconds	Total Liters of Water
2	90
4	180
6	270
8	360
10	

____

5. A particular water slide uses 5,000 gallons of water each month. Two water companies submitted plans to supply the water. Which company's plan would cost less per month?

Acme Water Company: $59.95 per month plus $0.02 per gallon of water

Best Water Supply: $19.95 per month plus $0.03 per gallon of water

____

6. The following table shows the height of the waves in a wave pool over 15 minutes.

Time (in minutes)	Wave Height (in feet)
3	$1\frac{7}{12}$
6	$2\frac{1}{12}$
9	$2\frac{7}{12}$
12	$3\frac{1}{12}$
15	$3\frac{7}{12}$

What is the pattern of the sequence? Write your answer in a complete sentence.

____

____

Subject Review

In this chapter, you worked with patterns and functions. A pattern consists of numbers that have something in common. A function is a relationship between an input and an output. You learned how to identify the rule of a pattern as well as how to calculate and predict the values of future numbers in a pattern. You also learned how to calculate the missing terms within an input-output table.

Here are the answers to the questions from page 87.

What blockbuster movie was initially rejected by Hollywood?

Amazingly, most Hollywood producers didn't think that making Star Wars *would have been a worthwhile pursuit. Instead, it changed the way movies were made forever.*

What meteor shower is visible every year in November?

The Leonid meteor shower has provided hours of amazement and entertainment for people all through history. It is one of nature's great free shows. It is usually visible in mid to late November.

How much time can it take to drop 85 feet on a water slide?

There is a large water park in Ohio called the Beach. At the Beach is a ride called the Banzai. If you ride it, you will fly down the triple-drop slide in only 6.5 seconds! Wow, that's some rush!

CHAPTER 11

Polygons and Solids

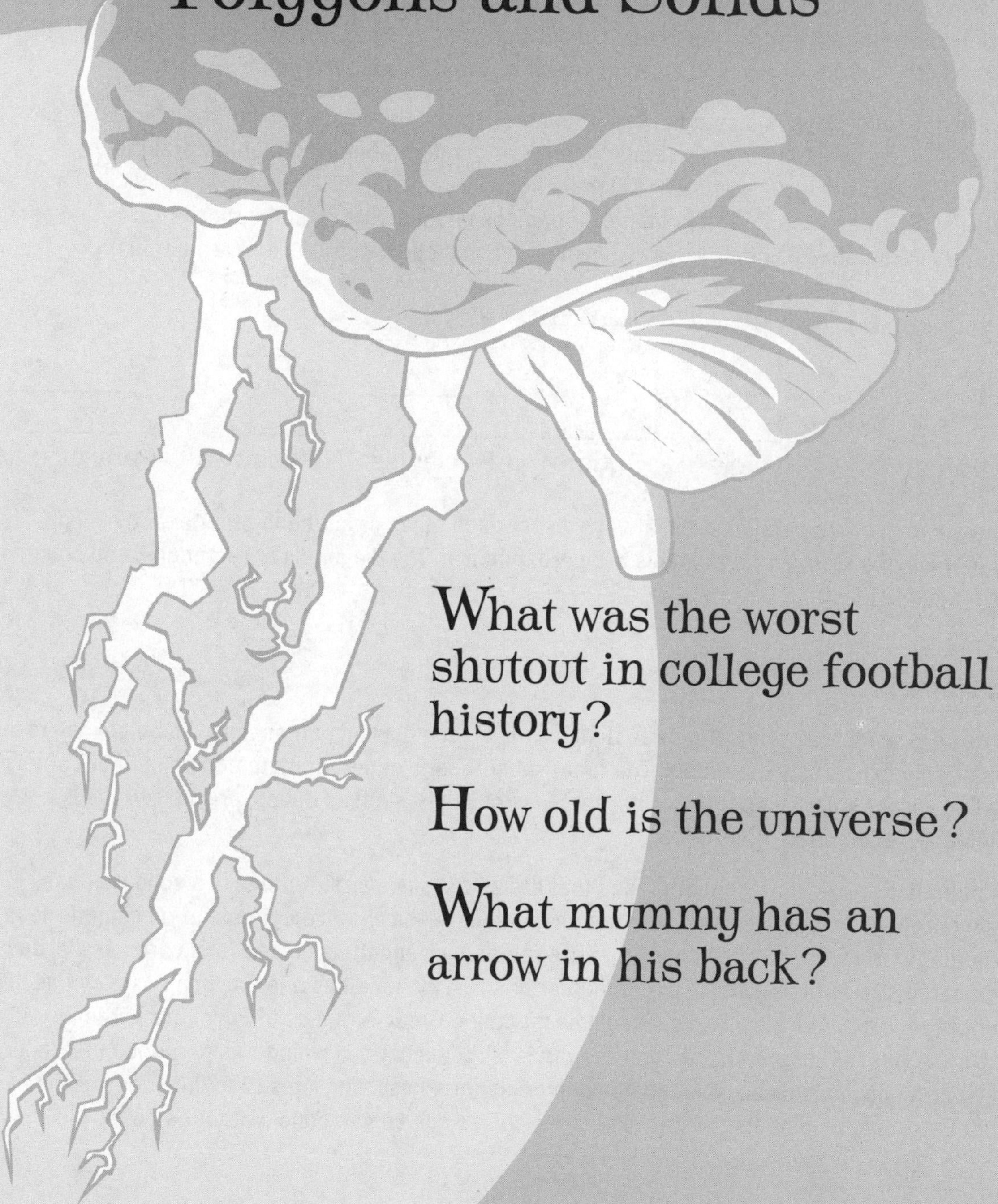

What was the worst shutout in college football history?

How old is the universe?

What mummy has an arrow in his back?

Polygons

A **polygon** is a closed, flat figure that is made up of line segments. Polygons can have any number of sides. A polygon with three sides is a **triangle.** Four-sided polygons are called **quadrilaterals.** A polygon with five sides is a **pentagon,** a polygon with six sides is a **hexagon,** a polygon with seven sides is a **heptagon,** and a polygon with eight sides is an **octagon.**

There are several types of quadrilaterals, such as rectangles, squares, and rhombuses. These types of quadrilaterals are also parallelograms. A **parallelogram** is a quadrilateral that has opposite sides that are parallel. A **rectangle** has four right angles and opposite sides that are parallel. **Squares** have four right angles, four sides of equal length, and opposite sides that are parallel. A **rhombus** has four sides of equal length and opposite sides that are parallel.

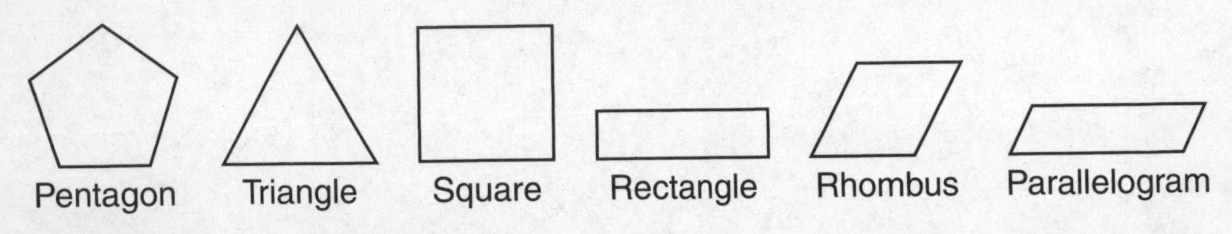

The diagonals of a parallelogram always bisect each other, or cut each other exactly in half. The diagonals of a square or a rhombus are perpendicular. The diagonals of rectangles and squares are equal in length.

Solids

A **solid** is a figure that has the three dimensions of length, width, and height. Some solids are made up of only flat surfaces, or **faces.** The faces meet to form edges, and the edges meet at corners called **vertices.** There are many different types of solids, such as cubes, prisms, pyramids, cylinders, and cones.

A **cube** is a solid with six square faces that are all the same size. A **prism** is a solid that has two parallel bases of the same size and shape and that has parallelograms for all its other faces. **Rectangular prisms** have six faces that are rectangles, and the opposite faces are parallel and the same size. A cereal box is a rectangular prism. A **pyramid** has one polygon base, and all the other faces are triangles. As you would suspect, the Great Pyramids of Egypt are pyramids. A **cylinder** has two parallel bases that are circles. The height of a cylinder is perpendicular to its bases. A can is a cylinder. A **cone** has one circle for a base, and lines from the edges of the base meet at a single point. Ice cream sometimes comes served in a cone (without a base).

When you look at a solid, it isn't possible to see all of the faces at once. You have to visualize the unseen faces. Below is a diagram of a rectangular prism to show you how one looks when it is drawn.

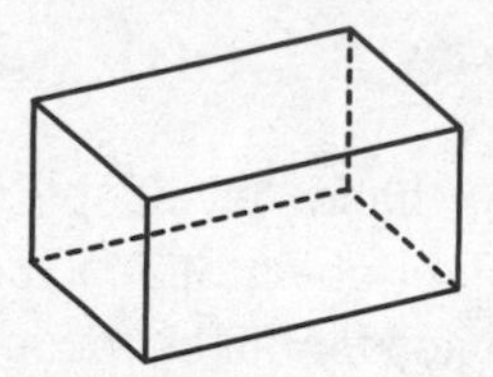

If you have trouble when answering a question about a three-dimensional solid, think about the shapes that make up the solid. Usually, the shapes are circles, squares, rectangles, or triangles.

Directions: Read the passage below and answer the question that follows.

The Biggest Shutout Ever

Can you imagine losing a game, any game, 222 to 0? Yowser! That's what happened to the football team at Cumberland University on October 7, 1916, when it played against the mighty team from Georgia Tech. How bad was it? Every time Georgia Tech ran a play, it scored. The game ended up being stopped 15 minutes early as a mercy to Cumberland. Can you imagine what the final score would have been if they had played all the way to the end?

► A football field is rectangular in shape and has diagonals that are equal in length. On a separate sheet of paper, draw another polygon that has diagonals of equal length.

Know It All Approach

Read the question carefully and figure out the information you need to answer the question correctly. The words you want are "another polygon" and "diagonals of equal length." There are no calculations for this problem. You will need to draw a polygon that has diagonals of equal length. The polygons that have diagonals of equal length are the rectangle and the square. In order to answer this question correctly, you must draw a square. Draw neatly! You should also write the word *square* underneath your drawing so that it is clear that you meant to draw a square.

Directions: Read the passage below and answer the questions that follow.

The Universe's Coolest Telescope

Did you know that NASA has a telescope in space that has taken some amazing photographs? The Hubble Space Telescope rocketed into space in 1990. The telescope is named for the astronomer Edwin Hubble, who believed that the universe gets bigger and bigger every second. The telescope named for Hubble has helped prove this to be true. The telescope has sent back amazing pictures of the births and deaths of stars, explosive bursts from distant galaxies, and the massive, all-consuming center of many galaxies—black holes.

Closer to home, the telescope has viewed volcanoes on Venus and discovered new moons around Saturn. With the help of the Hubble, astronomers have been able to determine that the universe is 12 to 14 billion years old. Wow!

1. You can make a model of the Hubble Space Telescope. Start with a piece of 2-inch-wide pipe that is $6\frac{3}{4}$ inches long. Which of the following solids is the same as the shape of a pipe?

   A cone
   B cylinder
   C pyramid
   D prism

2. To create a model of each of the solar panels of the Hubble Space Telescope, you will need to cut posterboard into polygons that are $6\frac{1}{2}$ inches long by 2 inches wide, with all right angles. What is the shape of the solar panels on Hubble?

   A rhombus
   B square
   C rectangle
   D hexagon

3. The Hubble has taken pictures of many stars. Four of the stars in a constellation make up a polygon with four right angles and parallel opposite sides. Draw a polygon that fits this description.

4. Another polygon is part of the Pegasus constellation. It has four equal sides that are parallel and four right angles. Which of the following is part of Pegasus?

A rectangle
B square
C octagon
D rhombus

5. If the Hubble Space Telescope passed over Egypt, what would the view of the top of the Great Pyramid look like? Hint: The great pyramid is a rectangular pyramid.

A 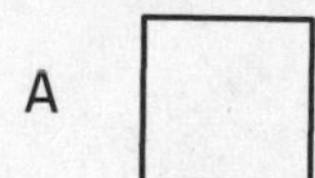

B 

C 

D

Directions: Read the passage below and answer the questions that follow.

Ötzi, the Leathery Mummy

A man's frozen body was found in the mountains between Austria and Italy in 1991. Tests showed that he had died about 5,000 years ago. The leathery mummy, given the name Ötzi for the area where he was found, had died with various tools of his time scattered around him: a bow and quiver, an axe, a flint dagger, and birchbark containers. Though it may not sound old to us now, his age was between 40 and 50 when he died, which was ancient 5,000 years ago. One way scientists were able to determine his age was from the signs of arthritis in his joints. There are tattoos on his knee, ankle, lower back, and one arm where he had arthritis. Some people think that Ötzi and people of that time may have thought that tattoos would lessen the pain of arthritis.

6. Ötzi carried a quiver that contained a dozen unfinished arrows. What would an end view of a finished arrow look like?

A 

B 

C 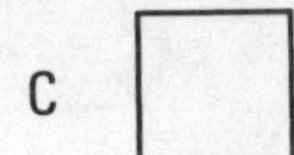

D 

7. Ötzi was high in the mountains when he died. If he had carried a tent for shelter and it had been shaped like a rectangular pyramid, what would be the shape of each side?

____

8. How many faces does a rectangular prism have?

____

9. Ötzi carried sloeberries for food, and his stomach contained bits of grain from bread. But it's doubtful Ötzi was ever lucky enough to eat a double-dip chocolate fudge ice cream cone with sprinkles! Draw a picture of a cone.

10. Another food item that Ötzi never saw was a sugar cube. How many equal faces does a sugar cube have?

    6 ______________________

11. How many vertices does a sugar cube have?

    ____

12. Some people believe that Ötzi came from a farming community. If a plot of land has four sides with parallel opposite sides and no right angles, which quadrilateral could this plot of land be?

    A rectangle
    B pentagon
    C rhombus
    D square

Subject Review

In this chapter, you learned about various types of polygons and shapes. The properties of rectangles, squares, rhombuses, and parallelograms were defined and illustrated. You also read that the diagonals of a square or rhombus are perpendicular and that those of a square or rectangle are equal in length.

Three-dimensional shapes such as cones, pyramids, cylinders, and prisms were also discussed. You know that three-dimensional shapes can have faces, edges, and vertices, and you now know the shapes of some three-dimensional objects in the real world.

Now that you've reviewed polygons and shapes, it's time to answer the questions from the beginning of the chapter.

What was the worst shutout in college football history?

The worst shutout in college football history occurred on October 7, 1916, when Cumberland University lost to Georgia Tech 222—0.

How old is the universe?

The Hubble Space Telescope has helped us determine that the universe is between 12 and 14 billion years old.

What mummy has an arrow in his back?

Ötzi, the Ice Man of the Alps, was found mummified and preserved in 1991. He had been well preserved in a glacier for more than 5,000 years.

CHAPTER 12

Area and Perimeter

What archeological finding in Denmark has puzzled historians for more than 100 years?

What teenage painter was a famous artist when she was only 10 years old?

Perimeter

The **perimeter** of a figure is the sum of the measures of all of its sides. The formula for the perimeter of a square is $P = 4s$. That's the same as $s + s + s + s$. If the side of a square is 13 inches long, the perimeter is $P = 4 \times 13$ inches, or 52 inches.

The formula to determine the perimeter of a rectangle or any parallelogram is $P = 2l + 2w$, where l is length and w is width. Suppose that the length of a rectangle is 10 yards and its width is 6 yards. $P = 2(10) + 2(6) = 20 + 12 = 32$ yards.

To find the perimeter of a triangle, add all three sides together. Suppose that a triangle's sides are 17, 9, and 12. $P = 17 + 9 + 12 = 38$.

Because a circle has no sides, it has no perimeter. The distance around a circle is called its **circumference.** The formula to determine the circumference of a circle is $C = 2\pi r$ (for radius) or $C = \pi d$ (for diameter). The value of π is about 3.14. As you may know, the radius is half the diameter, so you can use either formula. Suppose that the radius of a circle is 3 meters. The circumference is approximately 18.8 meters, because $C = 2\pi 3 \approx 18.8$ meters. Because the same circle has a diameter of 6 meters, $C = \pi 6 \approx 18.8$ meters. You can see that the formula works whether you have the radius or diameter measurement.

Area

The **area** is the amount of space covering the inside of a closed figure. It is measured in square units, such as square inches, square centimeters, square feet, or square meters.

The formula for the area of a square is $A = s^2$. Suppose that one side of a square is 9 inches. Because $A = (9 \text{ inches})^2$, the area of the square is 81 square inches. Another way to think of this is to imagine the square divided into smaller squares with each smaller square 1 inch long and 1 inch wide, like a grid. The square would be divided into 81 smaller squares.

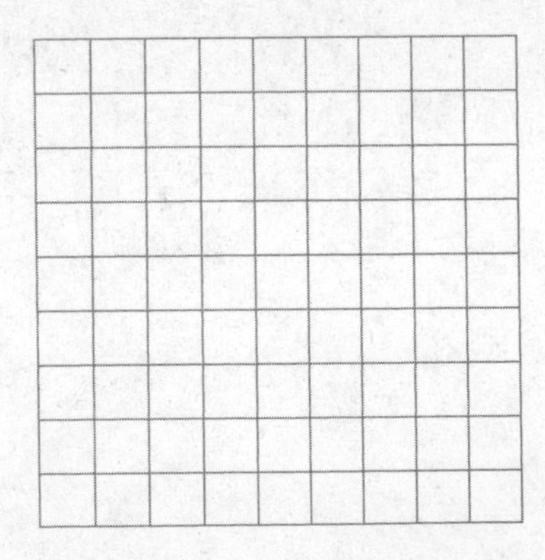

The formula for the area of a rectangle is $A = lw$. If the length of a rectangle is 5 meters and its width is 9 meters, then its area is $A = 5 \text{ m} \times 9 \text{ m}$, or 45 square meters. Dividing the rectangle into grid-like squares works here too.

The formula for the area of a parallelogram is $A = bh$. Because a parallelogram doesn't have a right angle, it does not have the same formula for determining the area of a rectangle. If you draw a line to make a right angle, you will get an altitude or height. If you cut off that segment and move it to the other side of the figure, you will have a rectangle.

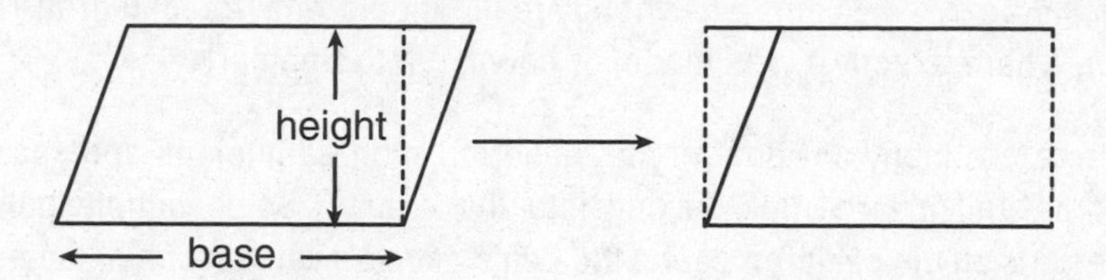

The formula for the area of a triangle is $A = \frac{1}{2}bh$. The reasoning behind this formula is that two triangles may form a parallelogram. Therefore, the formula to determine the area of a parallelogram is divided in half. If the base of a triangle is 12 centimeters and the height is 11 centimeters, the area of the triangle is $A = \frac{1}{2} \times 12 \text{ cm} \times 11 \text{ cm} = 66$ square centimeters.

The formula for the area of a circle is $A = \pi r^2$. Suppose that the radius of a circle is 9 feet. Then, $A = \pi \times (9 \text{ feet})^2 = \pi \times 81$ square feet ≈ 254 square feet. The answer is approximate because the value of π has been rounded to 3.14.

Be very careful to plug values into formulas correctly. Also, make sure to use the order of operations when solving problems. Solve any operations in parentheses first. Then, figure out the exponents. After that, solve multiplication and division from left to right. Finally, solve addition and subtraction from left to right.

Directions: Read the passage below and answer the question that follows.

Treasure of the Bog

Do you know what a cauldron is? A cauldron is a large kettle used for boiling. The Gundestrup Cauldron, made of silver around 100 B.C., was found extremely well preserved in a Denmark bog in 1891. A bog is an area of land similar to swamp or marsh. Nobody really knows what the cauldron was used for, where exactly it was made, or how it got to Denmark.

There are many designs on the cauldron, such as humans and snakes with horns and a large bull sinking into the ground. Some people believe the designs on the cauldron depict the Celtic people's version of hell. Some people believe that the cauldron was a religious vessel, and others say that it had sacrificial purposes. As with many historical artifacts, its true purpose may never be known.

▶ The Gundestrup Cauldron has a diameter of 69 centimeters. What is the approximate circumference of a circle with a diameter of 69 centimeters?

A 179 centimeters
B 217 centimeters
C 222 centimeters
D 236 centimeters

Know It All Approach

Read the question carefully and notice the information you need to solve the problem. The term "approximate circumference" and the number 69 are the information you should note. To calculate the answer, you'll want to use the formula for circumference with the diameter measure: $A = \pi d$.

Plug in the diameter measure and multiply by 3.14 to get $3.14(69) = 216.66$ centimeters. One way to double-check your answer would be to take 216.66 and divide it by 3.14. If you do this, you will get the answer 69. Because 69 is the diameter of the cauldron, you know you've calculated correctly.

When you calculated the answer, you may not have gotten a number that is in the list of answer choices. That's okay! The question asked for an *approximate* number. To find the correct answer choice, find the one that is approximately 216.66.

Use Process of Elimination to eliminate answer choices that are not near 216.66. When you round 216.66 to the nearest whole number, you get 217. You can eliminate answer choice (A) because it is too small. You can eliminate answer choices (C) and (D) because they are too large. Answer choice (B), 217 centimeters, is exactly (well, approximately) right.

Directions: Read the passage below and answer the questions that follow.

Olivia Bennett: World-Famous Teenage Painter

Olivia Bennett is an amazing painter who in 2003, at the age of 13, was named "One of Twenty Teens Who Will Change the World." But what's most amazing about her is how she learned to paint. Bennett started using a paintbrush to help recover from leukemia. Bennett loved painting. In fact, her creations were so terrific that they have been compared to the works of famous artists like Georgia O'Keefe. As cheerful as Bennett's paintings are, her story is just as joyful. Bennett has recovered well from leukemia and has served as an inspiration for many others. In 2002, a book called *A Life in Full Bloom* was published with 40 of Bennett's paintings. Bennett's description of her inspiration for each of painting is included alongside her work. Bennett has appeared in such magazines as *Teen People* and on the talk show *The Oprah Winfrey Show.*

1. **Part A**

   One of Olivia Bennett's paintings is of a dark purple flower called a hydrangea. The canvas measures 22 inches by 36 inches. What is the area of the painting?

   ____

   Part B

   What is the perimeter of the painting?

   ____

2. Suppose that the painting was placed in a frame that was 3 inches longer and 3 inches wider than the painting. How would the perimeter of the frame be different from the perimeter of the painting?

   A The perimeter would increase by 6 inches.
   B The perimeter would triple.
   C The perimeter would increase by 12 inches.
   D The perimeter would quadruple.

3. A painting that Olivia Bennett created at age 10 is a square, 64-square-inch watercolor titled "Small Autumn Leaves." If Bennett wanted to paint it again and double the length of each side, how many times larger will the new painting be?

   A 3 times the size of the original
   B 6 times the size of the original
   C 8 times the size of the original
   D 4 times the size of the original

4. If Olivia Bennett painted on a triangular canvas that had a base of 14 inches and a height of 12 inches, what would the area of the painting be?

   ____

5. If Bennett decided to paint a flower on a circular canvas with a radius of 11 inches, what would the approximate area of the painting be?

   A 379.9 square inches
   B 452.2 square inches
   C 296.3 square inches
   D 355.6 square inches

6. What is the approximate circumference of a circle with a radius of 11 inches?

A 34.5 inches
B 69.1 inches
C 47.7 inches
D 379.9 inches

7. **Part A**

The floor of an artist's studio is shown below. What is the perimeter of the floor?

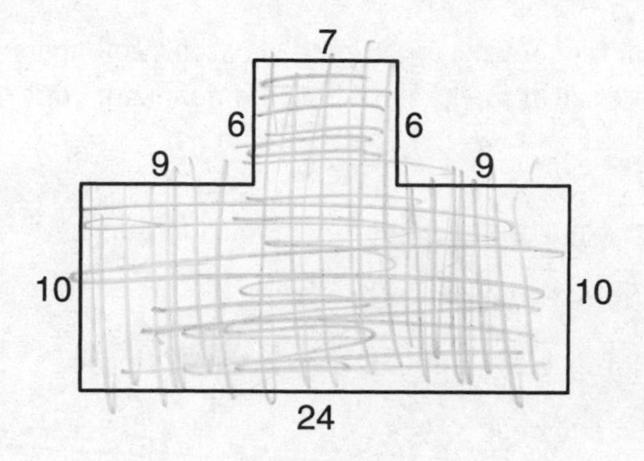

____

Part B

The floor of the artist's studio is going to be tiled. The person laying the tile will need to know the area of the floor to know how much tile to use. What is the area of the studio floor?

____

Subject Review

Phew! There were a lot of formulas in this chapter. You learned how to define the perimeter and area for various shapes and objects. You worked with the perimeter formula for squares, rectangles, and triangles and the circumference formula for a circle. Remember, the perimeter of an object is the measurement of all of its sides added together.

The area of an object is a little more complicated. The area of an object is the measurement of the space contained within the object.

You worked with the formula for determining area of squares, rectangles, parallelograms, triangles, and circles. Do you remember all of them? Here they are again, for your reference.

The formula for the perimeter of a square is $P = 4s$.

The formula for the perimeter of a rectangle or any parallelogram is $P = 2l + 2w$.

The formula for the perimeter of a triangle is $P = 3s$.

The formula for the circumference of a circle is $C = 2\pi r$ or $C = \pi d$.

The formula for the area of a square is $A = s^2$.

The formula for the area of a rectangle is $A = lw$.

The formula for the area of a parallelogram is $A = bh$.

The formula for the area of a triangle is $A = \frac{1}{2}bh$

The formula for the area of a circle is $A = \pi r^2$.

And now, the answers you've been waiting for:

What archeological finding in Denmark has puzzled historians for more than a hundred years?

Historians still debate exactly who made the Gundestrup Cauldron. It is not even known precisely what the cauldron was used for.

What teenage painter was a famous artist when she was only 10 years old?

Olivia Bennett's work has proven that people of any age can create beautiful art.

CHAPTER 13

Surface Area and Volume

How far can a 140-pound barrel be rolled in one day?

Where could you see a coffin shaped like a fish?

Surface Area

The **surface area** of a three-dimensional figure is the sum of the areas of all of its surfaces.

The formula for the surface area of a rectangular prism is $S = 2lh + 2lw + 2wh$, where l is the length, h is the height, and w is the width of the prism. The formula adds the area of all six faces.

Suppose a shirt box is 14 inches long, 3 inches high, and 9 inches wide. $S = 2(14 \text{ in.})(3 \text{ in.}) + 2(14 \text{ in.})(9 \text{ in.}) + 2(9 \text{ in.})(3 \text{ in.}) = 84 \text{ in.}^2 + 252 \text{ in.}^2 + 54 \text{ in.}^2 = 390 \text{ in.}^2$. The shirt box has a surface area of 390 square inches.

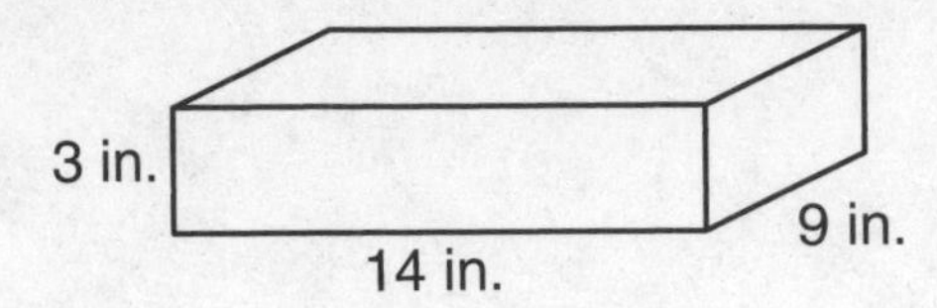

The formula for the surface area of a cylinder is $S = 2\pi r^2 + 2\pi rh$, where r is the radius and h is the height of the cylinder. This formula adds the area of the top and bottom bases of the cylinder to the area of the curved surface, which is a rectangle when flat.

Suppose a can of soup has a radius of 8 centimeters and a height of 13 centimeters. $S = 2\pi(8 \text{ cm})^2 + 2\pi(8 \text{ cm})(13 \text{ cm}) = 2\pi 64 \text{ cm}^2 + 2\pi 104 \text{ cm}^2 = 401.92 \text{ cm}^2 + 653.12 \text{ cm}^2 = 1{,}055.04 \text{ cm}^2$. The soup can has a surface area of 1,055.04 square centimeters.

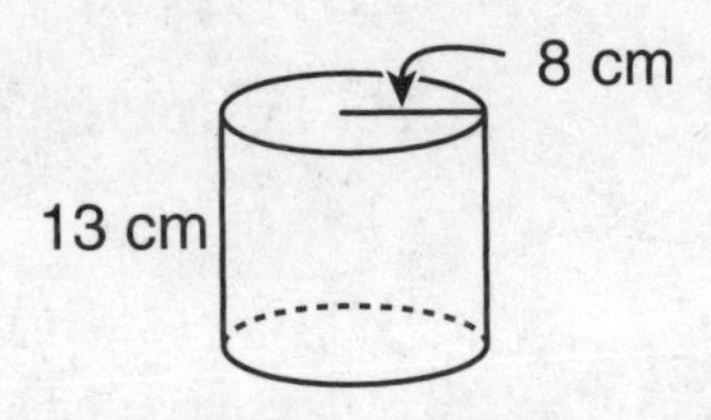

Volume

Volume is the amount of space inside a three-dimensional figure. For example, the amount of space that could fit in a bathtub would be called volume.

The formula for the volume of a rectangular prism is $V = lwh$. If a box of cereal is 6 inches long, 2 inches wide, and 9 inches high, its volume is $V = (6 \text{ in.})(2 \text{ in.})(9 \text{ in.}) = 108$ cubic inches.

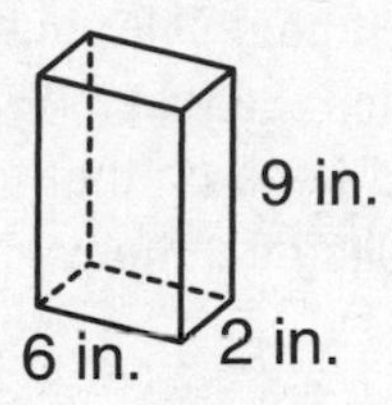

The formula for the volume of a pyramid is $V = \frac{1}{3}Bh$, where B is the area of the base and h is the height. The base of a pyramid can be a triangle, a square, or a multisided polygon. A given pyramid has a square base with sides of 6 meters and a height of 8 meters. For this figure, the formula is $V = \frac{1}{3}(s^2)(h)$. $V = \frac{1}{3}(6\text{m})^2(8 \text{ m}) = \frac{1}{3}(288\text{m}^3) = 96$ cubic meters.

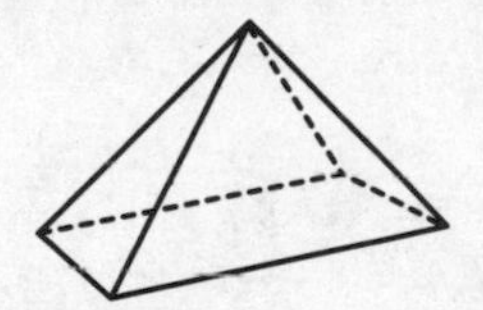

rectangular pyramid

The formula for the volume of a cylinder is $V = \pi r^2 h$. Chapter 12 said that πr^2 is the formula for the area of a circle. That area is multiplied by the height of the cylinder to give you the volume. Suppose an oatmeal box has a radius of 7 centimeters and a height of 23 centimeters. $V = \pi(7 \text{ cm})^2(23 \text{ cm}) = \pi(49 \text{ cm}^2)(23 \text{ cm}) = 3{,}538.78 \text{ cm}^3$. The volume is approximately 3,539 cubic centimeters.

Directions: Read the passage below and answer the question that follows.

Rolling, Rolling, Rolling

Some people will go way out of their way just to get a good workout. Within a 24-hour period in 1998, a team from the Netherlands rolled a barrel that weighed 63.5 kilograms (about 140 pounds) a distance of 263.9 kilometers (farther than the distance between Washington, D.C., and Philadelphia, Pennsylvania). If you had been on that team, wouldn't you have hoped that your turn consisted of rolling that heavy barrel down a nice long hill?

▶ If the height of a barrel is 90 centimeters and its diameter is 40 centimeters, what is its surface area?

A 32,656 cubic centimeters
B 9,834 cubic centimeters
C 13,816 cubic centimeters
D 27,274 cubic centimeters

Know It All Approach

Read the question carefully and take note of the information you'll need. The numbers 90 and 40 and the terms "diameter" and "surface area" are the information needed to solve it. The formula for the surface area of a cylinder is $S = 2\pi r^2 + 2\pi rh$.

The tricky part of the question is that it gives you the diameter of the barrel but you need the radius for the formula. Don't worry! The radius is half the diameter, or 20 centimeters. Now, plug the measurements into the formula.

$S = 2(3.14)(20 \text{ cm})^2 + 2(3.14)(20 \text{ cm})(90 \text{ cm})$

$S = (6.28)(400 \text{ cm}^2) + (6.28)(1{,}800 \text{ cm}^2)$

$S = 2{,}512 \text{ cm}^2 + 11{,}304 \text{ cm}^2 = 13{,}816 \text{ cm}^2$

To double-check your answer, make sure that you have written the right numbers from the question. Then, look to see that they are plugged into the formula correctly. Follow the order of operations and calculate the answer again. Answer choice (C) is the same as your calculation and is therefore correct.

Directions: Read the passage below and answer the questions that follow.

Ghana's Decorative Coffins

In the United States, coffins are usually serious and dignified, but they are seen in a different way in the African country of Ghana. It is becoming a tradition there to create beautiful, colorful coffins shaped like favorite objects in the life of the departed. A fisherman, for example, was buried in a fish-shaped coffin. A man who loved Mercedes-Benz automobiles was buried in a casket shaped like the cars. A hunter had a coffin that resembled a leopard.

Expert woodworkers carve the coffins requested by the family members of the deceased. It may take a month or longer to carve and paint the detailed designs. The coffins can be quite expensive. In some cases, they cost as much as a Ghanaian may make in a year. However, the coffins are viewed as part of a celebration of the life of a loved one, and to many people, such a celebration comes without concern for the price tag.

1. Suppose that a farmer chose a coffin shaped as a green onion. If the dimensions of the cylinder are 6.5 feet high and 4 feet in diameter, what is the approximate surface area of the coffin?

   A 118 square feet
   B 107 square feet
   C 264 square feet
   D 68 square feet

2. If a rectangular coffin measures 2 meters long, 1.5 meters wide, and 1 meter high, describe a way to find the volume of the coffin.

   ____

   ____

   ____

3. One coffin was made to look like a truck for hauling coal. The bed of the truck contained painted blocks of wooden coal. Suppose that each block was 1 cubic foot and the bed of the truck measured 8 feet by 4 feet by 3 feet. How many blocks were there?

_______________________ blocks

4. In Egypt, pyramids were used to bury the dead. If someone in Ghana wanted to be buried in a much smaller pyramid that had a base of 2.5 meters by 2 meters and a height of 3 meters, what would be the volume of the pyramid?

   A 15 cubic meters
   B 9 cubic meters
   C 18 cubic meters
   D 5 cubic meters

5. Thousands of people may attend a funeral in Ghana. Posts with rope may be used to keep cars out of certain areas. Suppose that some of the cylindrical posts have a radius of 15 centimeters and a height of 60 centimeters and others have a radius of 15 centimeters and a height of 30 centimeters. How is the volume of the larger post different from the volume of the smaller post?

   A The volume is decreased by 2.
   B The volume is three times as large.
   C The volume is twice as large.
   D The volume is half as large.

6. An artist creates two wooden models of rectangular coffins as a way to advertise her work. Model A measures 12 inches by 9 inches by 3 inches. Model B measures 24 inches by 18 inches by 6 inches.

Part A

How much more surface area does Model B have than Model A?

____

Part B

How much more volume does Model B have than Model A?

____

7. The volume of a crate used to transport a coffin is 960 cubic feet. The height of the crate is 8 feet. What could be the length and width of the crate?

A 9 feet by 6 feet
B 11 feet by 8 feet
C 10 feet by 12 feet
D 14 feet by 11 feet

8. On a trip to Ghana, a museum director purchased many pieces of art that she wanted to transport back to the museum. One will go into a box, and another will go into a plastic cylinder. The box and the cylinder have almost the same volume. Is the surface area also almost the same? If not, what is the difference?

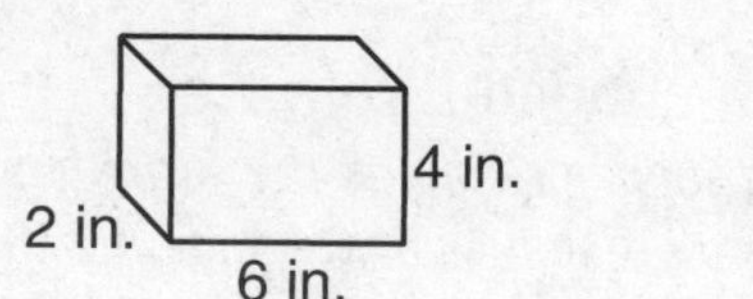

____

____

Subject Review

In this chapter, you learned the formulas for the surface area of a rectangular prism and of a cylinder. You also learned the formulas for the volume of a rectangular prism, a pyramid, and a cylinder. Here they are again, for your reference.

The formula for the surface area of a rectangular prism is $S = 2lh + 2lw + 2wh$.

The formula for the surface area of a cylinder is $S = 2\pi r^2 + 2\pi rh$.

The formula for the volume of rectangular prism is $V = lwh$.

The formula for the volume of a pyramid is $V = \frac{1}{3}Bh$.

The formula for the volume of a cylinder is $V = \pi r^2 h$.

Remember that surface area is the sum of the areas of a three-dimensional figure's surfaces. For example, the surface area of a gift box may be the same as the area of the wrapping paper used to wrap it. Remember that volume is the amount of space within a three-dimensional figure. For example, the volume of a can of soda pop is the same as the amount of soda inside of it.

The main thing to remember is to take your time when you plug numbers into a formula. Some formulas use both addition and multiplication, so be sure to get the operations right and to perform your calculations in the correct order.

The answers to the questions on page 113 are below.

How far can a 140-pound barrel be rolled in one day?

A team from the Netherlands managed to roll a 140-pound barrel 164 miles in a single day. Think you and your friends can do better?

Where could you see a coffin shaped like a fish?

You can see coffins of just about any shape or color in the African country of Ghana, where exquisitely designed coffins are created as a way to honor the deceased.

CHAPTER 14

Similarity and Congruence

Who made the world's largest flag?

Before the Roman gladiators fought lions, who did they fight against?

Similarity

Similar figures are figures that are the same shape and also have corresponding angles that are equal. Their sides may be of different lengths, but they are in proportion to each other. Figures *A* and *B* below are similar, and Figures *X* and *Z* are similar. The two sets of figures have sides that are in proportion to each other and have corresponding angles that are equal.

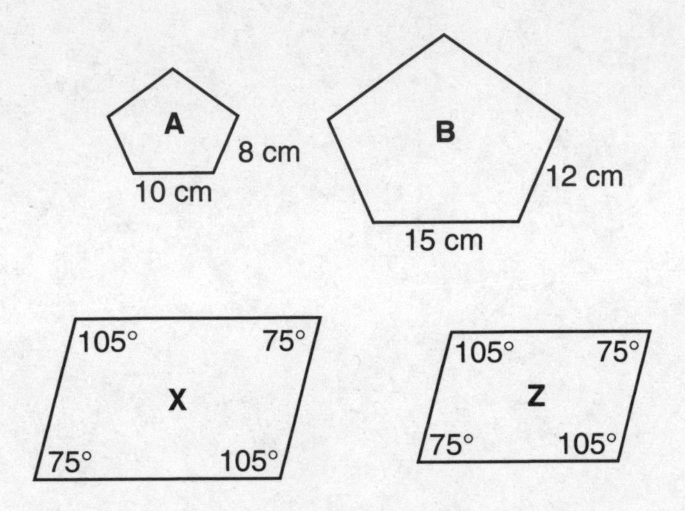

If two figures are similar, they will match up exactly if one of them is magnified or shrunk. (You may have to turn one of the figures around, though.)

Congruence

Congruent figures are figures with the same size and shape. Figures *M* and *N* below are congruent because they are exactly the same.

2 in. 2 in. *M* 2 in. 2 in.

2 in. 2 in. *N* 2 in. 2 in.

Congruent angles are angles with identical measures. Angles *F* and *G* below are congruent because they are exactly the same size.

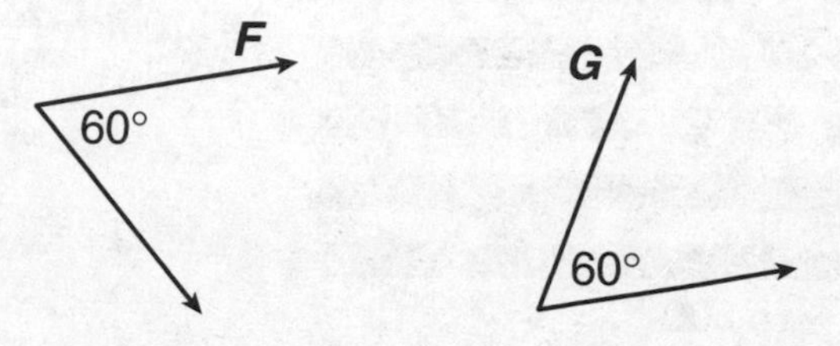

If two figures are congruent, they will match up exactly if one is placed on top of the other.

Directions: Read the passage below and answer the question that follows.

In 1996, Thomas "Ski" Demski had his "Superflag" suspended across the Hoover Dam to show how large it was. Demski's American flag weighs 3,000 pounds, which is about the weight of a car. It is large enough to cover an entire football field. The flag has been used at sporting events like the Super Bowl and the World Series, as well as patriotic events. Mr. Demski said that the purpose of a flag that large is to help bring about feelings of patriotism in our society.

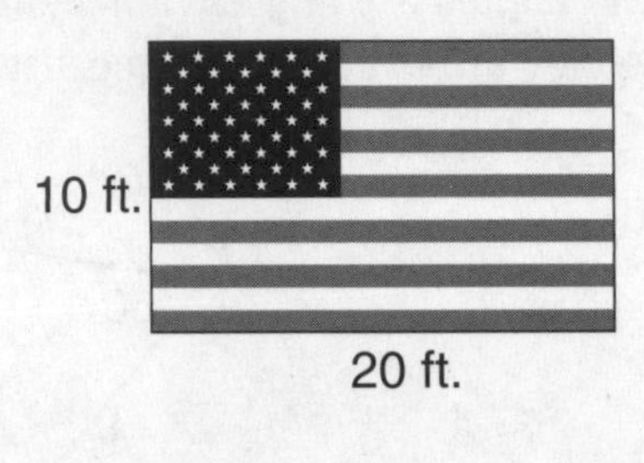

► Look at the two flags above. How can you tell if they are similar?

____

____

____

Know It All Approach

Read the question carefully and find the information you'll need to answer it. The most valuable word is "similar." Use the space provided to calculate your answer. Write neatly and show all of your work.

In this particular case, you need to determine what the relationship is between the two flags. Compare the ratios of corresponding sides to see if they are in proportion to each other. $\frac{20}{500} = \frac{1}{25}$ and $\frac{10}{250} = \frac{1}{25}$. The ratios of the corresponding lengths are equal, so the two flags are similar.

Your written answer should be something like: *The two flags are similar because the ratios of their corresponding lengths is equal and the flags are in proportion.* Now, you should double-check your answer. Make sure that you have used the correct numbers from the question and have worked out the proportion properly. Make sure that your answer can be easily read and that you have neatly written the work you did to find the answer.

Directions: Read the passage below and answer the questions that follow.

The Life of a Gladiator

The movie *Gladiator* showed the difficult times of one Roman gladiator. The life of an actual gladiator was a bit different. When the contest started in 264 B.C., it featured people fighting people. It became a popular spectator sport very quickly. The gladiators were prisoners of war, slaves, and criminals, but in a sense they became celebrities. They trained in special schools to fight well, using a sword, dagger, trident, chain, and other lethal weapons. They were well fed, received good medical care, and sometimes had families. They could win their freedom by being popular with the crowd or winning bonus competitions, but they could never be citizens. Later, other spectacles were added to the gladiatorial games in the form of wild animals like lions and elephants.

Directions: Use the illustration below to answer questions 1 and 2.

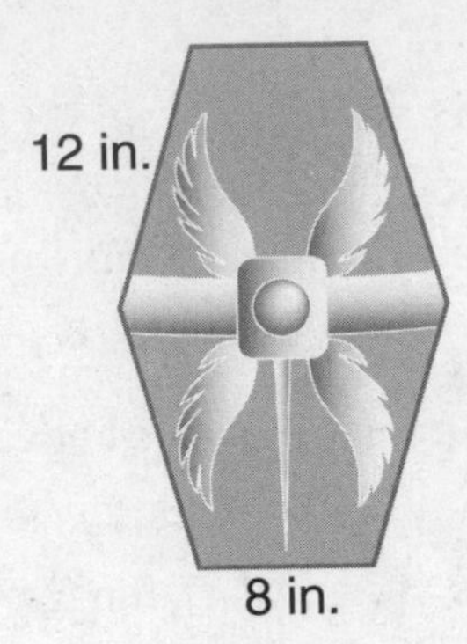

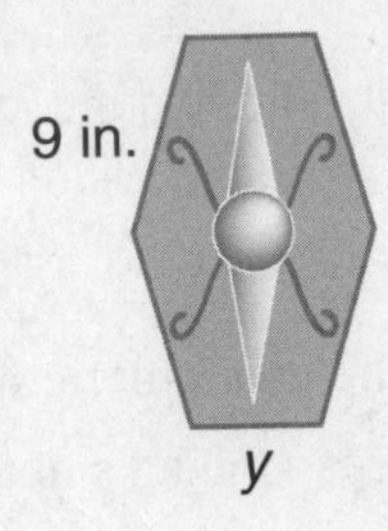

1. The gladiator shields above were made for men of different heights. The shapes of the shields are similar. What is the value of y?

   A 9 inches
   B 4 inches
   C 6 inches
   D 3 inches

2. Are the shields above congruent? Explain why or why not.

____

____

____

3. The rectangle below represents a carpenter's design for a gladiator shield. Draw a design for a similar shield that has sides 2 times the length and width of the one shown.

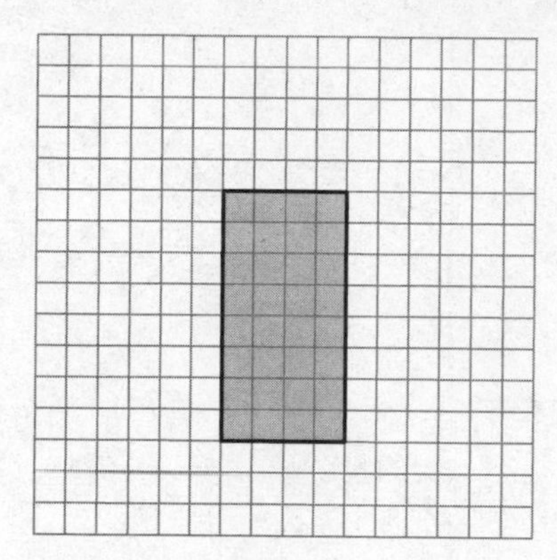

4. Look at the shields below. Which ones are congruent?

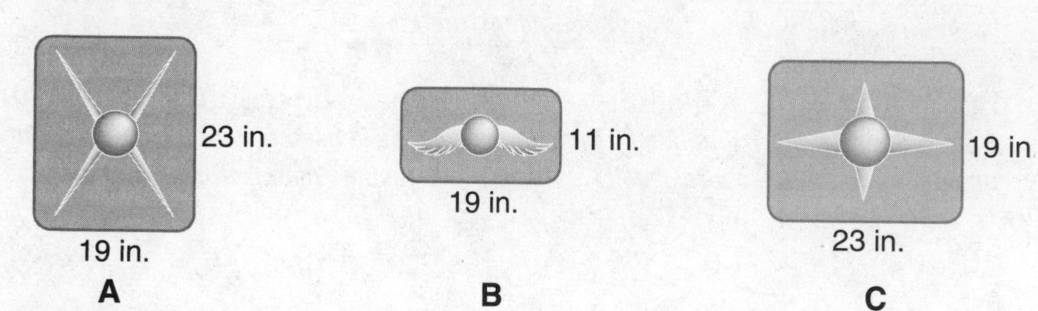

A A and B
B A and C
C B and C
D none of them

5. A gladiator is making a play shield for his son. He wants it to look like the one he uses. If his has the dimensions of the shield below, what will be the missing measurement for his son's shield?

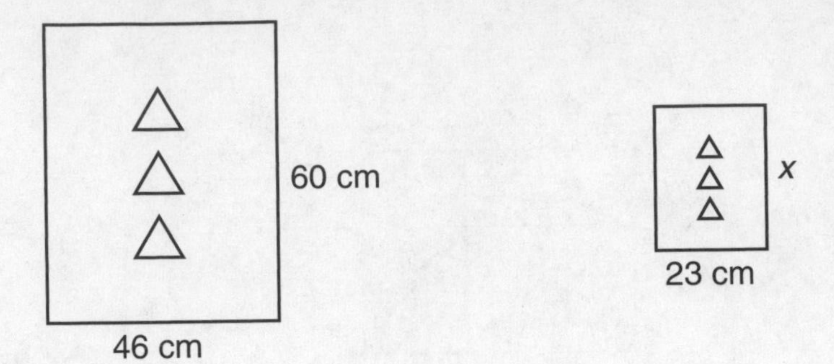

A 35 centimeters
B 20 centimeters
C 46 centimeters
D 30 centimeters

Subject Review

In this chapter, you learned about similar and congruent figures. Similar figures have corresponding angles that are the same size and corresponding sides that are in proportion to each other. Congruent angles and figures are exactly the same.

Who made the world's largest flag?

The American patriot "Ski" Demski created the world's largest flag. So far, a flagpole large enough to display it has not been constructed.

Before the Roman gladiators fought lions, who did they fight against?

Before the gladiators fought against animals such as lions, they were forced to fight against each other. Losers usually lost more than just the match. They lost their lives.

CHAPTER 15
Transformations

What bandit wore a bucket on his head?

Why did Leonardo da Vinci write backward?

Transformations

A geometric figure can be changed in many ways. A figure can be stretched or shrunk, spun around, or moved. When a geometric figure moves, it is called a **transformation.** Following are three types of transformations.

Reflections

A **reflection** is the mirror image of a geometric figure. Another name for it is a flip, because the figure is flipped across an invisible line.

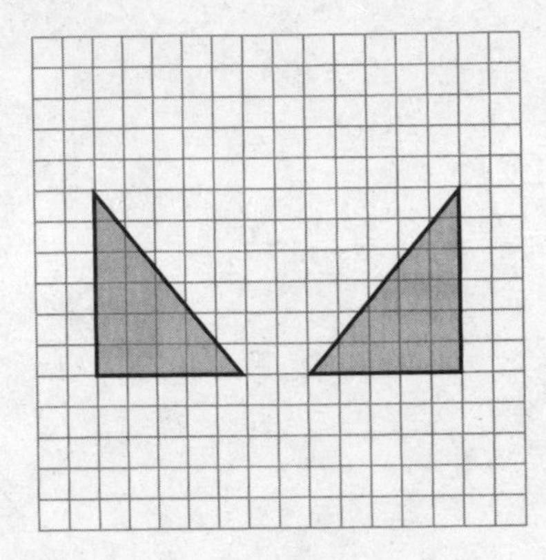

Rotations

A **rotation** turns a geometric figure around a fixed point on the figure. This is also called a turn.

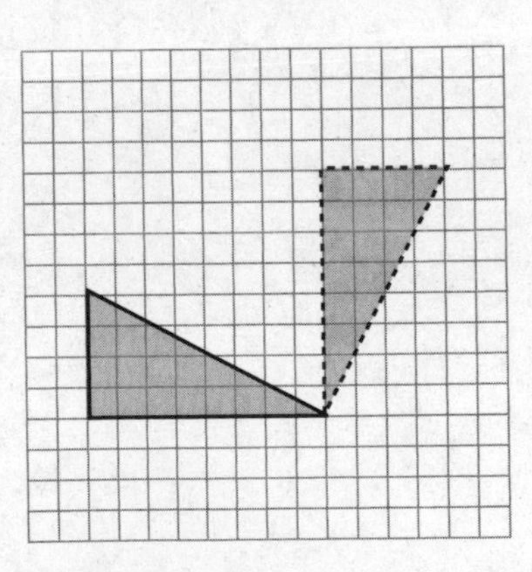

Translations

A **translation** is a change to the position of a geometric figure. This is also called a slide, because it's like moving an object as if it were sliding.

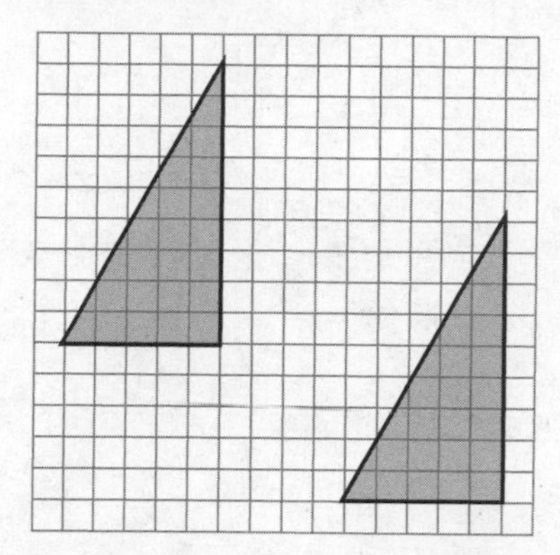

Some shapes have a **line of symmetry.** If a figure is folded along an invisible line and both halves are exactly the same, the figure has symmetry. There's also symmetry in transformations, because the figure, whether you flip it, turn it, or slide it, will never change shape. The figure following a reflection, rotation, or translation has the same size and shape as the original figure. Do you remember what we called figures that have the same size and shape? That's right! They're congruent.

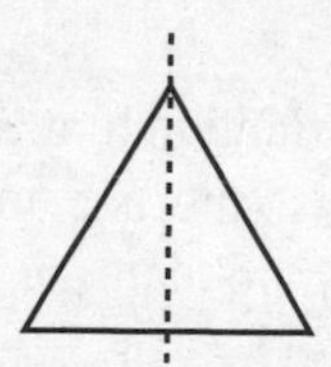

Directions: Read the passage below and answer the question that follows.

The Bucket-Head Bandit

Ned Kelly was considered a hero by some people and a criminal by others. In the 1800s in Australia, life was very hard and the government was run very poorly. Ned Kelly was 15 when he was arrested for riding a stolen horse. Though he didn't know the horse was stolen, he received a 3-year prison sentence. When he got out, the police had taken almost everything he owned. He formed a gang that committed robberies, but they only stole from the rich. To protect himself from gunfire, Ned Kelly wore a handmade suit of armor. He used a metal bucket to create a very bizarre helmet. It served its purpose until the police saw that there was no armor on Ned Kelly's legs. They shot his legs, and he was captured.

Which of the following describes the transformation between Helmet A and Helmet B?

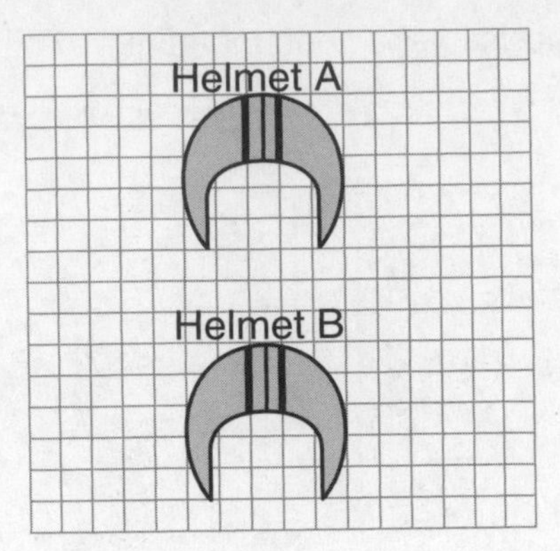

A rotation
B reflection
C translation
D there was no transformation

Know It All Approach

Read the question carefully and find the information that will get you the answer. The words to watch are "transformation between." Because there is no calculation required for this question, you will need to make a close observation of the diagram. How is Helmet A different from Helmet B?

You can see that they are the same figure except one is higher on the grid than the other. When the positions of figures change, a translation occurs. To double-check your answer, think of the definitions of a rotation and a reflection. The two helmets don't satisfy either of those definitions, so your answer is correct. You can eliminate answer choice (D) because it is clear that there is a difference. One helmet is in a different place on the grid than the other. You can eliminate answer choice (A) because if the images were rotations, they would be right next to one another. Choice (C) is correct.

To discover how figures have moved, draw an arrow from the angles and sides of one figure to the corresponding angles or sides of the other. Then, think about what movement was used to get the figure from one place to the other.

Directions: Read the passage below and answer the questions that follow.

Leonardo the Great

You may have heard of Leonardo da Vinci because he painted one of the most famous paintings in the world, the Mona Lisa. (There she is to the left!) However, da Vinci actually completed very few paintings. Instead, he designed fortresses, churches, and canals. His drawings include designs for missiles, tanks, flying machines, grenades, and machine guns. There's even a drawing of a submarine. What's so impressive is that he made these drawings before the year 1500, before anyone else had even thought of such inventions!

There's a legend that when da Vinci was a teenager, his father asked him to paint a shield. Well, good old Leonardo didn't think it would be cool to draw something ordinary, so instead he used gross-looking animals, like bats and maggots and lizards, as models. His finished painting for the shield was of a creepy monster spewing smoke and poison gas. Awesome!

1. **Part A**

   Leonardo da Vinci wrote backwards, from right to left. People use mirrors to read the detailed notes he left. What is the transformation on the first letter of Leonardo's name?

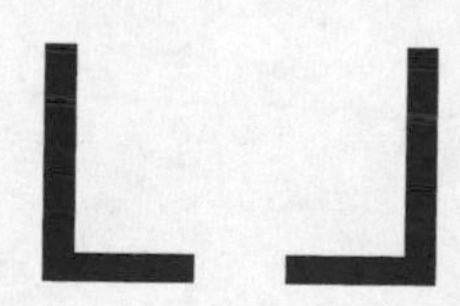

____

Part B

Describe how the letter was transformed.

flip

____

____

2. **Part A**

Leonardo da Vinci made many sketches. If he had made the arrow below, draw what it would look like if it were rotated.

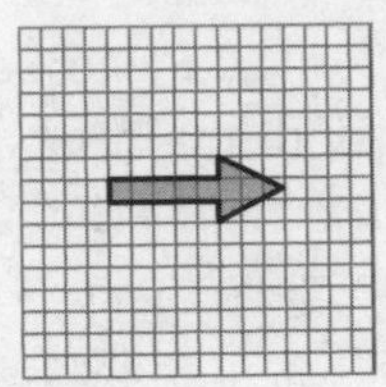

Part B

Describe how you rotated the arrow.

____

____

3. Which term best describes how the figure below was transformed?

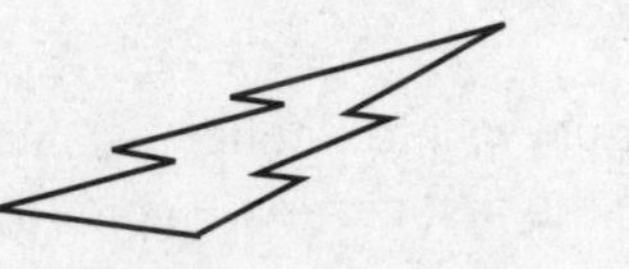

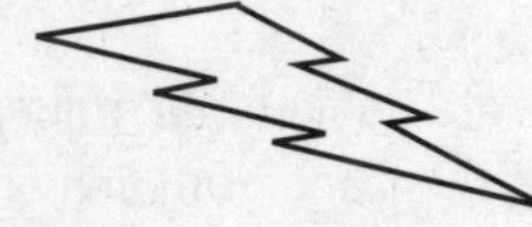

A reflection then rotation
B translation then reflection
C reflection then translation
D translation then rotation

4. Figure B is a transformation of Figure A.

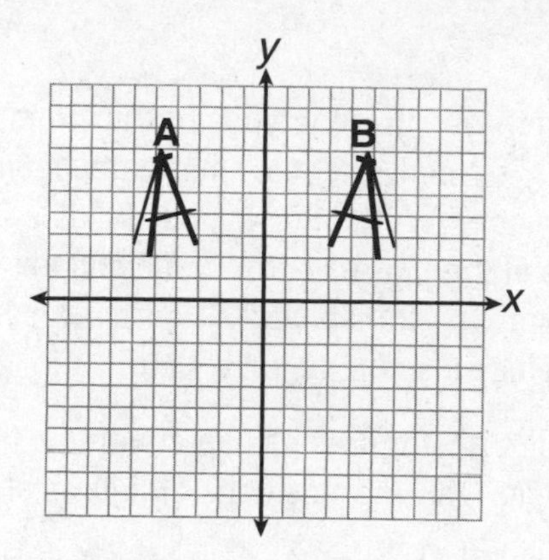

Use the lines below to name and describe the transformation.

____

____

5. Which term best describes how the figure below was transformed?

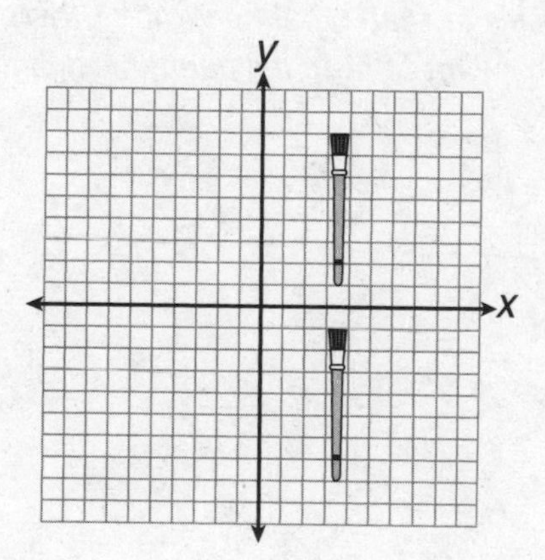

A turn
B reflection
C translation
D rotation

Subject Review

In this chapter, you learned about three different types of transformations: reflections, rotations, and translations. A transformation is simply the movement of a geometric figure. Remember that a figure's reflection is its mirror image. A reflection is also called a flip. During a rotation, two figures meet at a fixed point. One of the figures is rotated along this point, like the hand of a clock. A rotation can also be called a turn. In a translation, a figure moves away from an original figure in a sliding manner. A translation may also be called a slide.

A key point to remember is that all transformations use congruent figures, meaning that the shape and size of the figures don't change. Only the positions of figures change during transformations.

Here are the answers to the questions from the beginning of the chapter.

What bandit wore a bucket on his head?

Ned Kelly, the Robin Hood of Australia, wore a bucket on his head as part of a nearly full set of homemade armor. Too bad he didn't wear armor on his legs!

Why did Leonardo da Vinci write backward?

There are a couple of theories about why da Vinci wrote backward. Some believe that he didn't want others to be able to read what he wrote. Others believe it was because he was left-handed and didn't want to drag his hand through the wet ink of what he had just written.

Directions: Read the passage below and answer the questions that follow.

Maggots: The Larvae of Flies

Have you ever seen creeping, crawling, slimy, small white worms in old, rotting garbage? Those are maggots, the babies of flies. As gross as they look, they also serve some valuable purposes. One of their purposes is to help rotting flesh break down more quickly. Maggots eat away at the flesh of dead animals that scavengers have left behind. Another useful purpose for maggots is that because they eat dead tissue, but not healthy tissue, they help eliminate harmful bacteria in open wounds. That's right—maggots are used by doctors to eat away diseased flesh on people. Eeuuuw!

1. If 5,000 maggots are delivered equally to a group of 200 hospitals every week, which equation will show how many are delivered to each hospital?

   A $5{,}000 = x + 200$

   B $200 = 5{,}000x$

   C $5{,}000 = 200x$

   D $200 = 5{,}000 - x$

2. A female housefly can lay eggs up to seven times in her short lifespan. The eggs take about 8 hours to hatch into maggots. The table below is the number of eggs a housefly laid in her life. What is the value of m (for maggot)?

Laying Time	Total Eggs
1	105
2	220
3	345
4	480
5	m

____

3. Maggots were shipped to a hospital in a container with air holes. Suppose that a container is a solid figure with 8 vertices, 6 faces, and 3 pairs of faces of equal size. What is the shape of the container used to ship maggots?

   A pyramid
   B rectangular prism
   C cone
   D cylinder

4. **Part A**

   Maggots are packed in sterile material before being placed on a wound. If the sterile material is a rectangle 7.2 centimeters long by 6.8 centimeters wide, what is the area of the material?

   ____

   Part B

   What is the perimeter of the sterile material?

   ____

5. **Part A**

   In one hospital, a patient needed maggots on different areas of his leg. There was a total of 750 maggots in 21 bags. One of the sterile bags contained 50 maggots, but the other 20 were a little smaller and all contained the same amount. Write an equation that would find how many maggots were in each of the other bags.

   ____

   Part B

   How many maggots were in each of the 20 bags?

   ____

6. When laid flat, two sterile bags are different sizes but are similar rectangles. What is the measurement of the missing side of the second figure below?

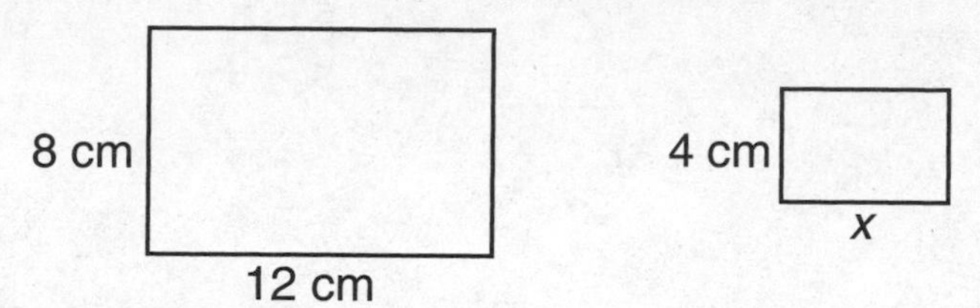

A 10 centimeters
B 4 centimeters
C 8 centimeters
D 6 centimeters

7. Pasquale packs maggots in sterile bags. He then carries them to the patient's room in a box. The box is 9 inches long, 6 inches wide, and 2 inches high. What is the surface area of the box?

A 168 square inches
B 164 square inches
C 152 square inches
D 158 square inches

8. A doctor applied a pyramid-shaped bandage to protect an injured area. The base of the pyramid, which was under the patient's leg, was a square that was 7 inches on each side. The height of the pyramid was 12 inches. What was the volume of the pyramid?

A 222 cubic inches
B 196 cubic inches
C 204 cubic inches
D 218 cubic inches

9. Which of the following is a possible view of one of the sides of the pyramid described in question 8?

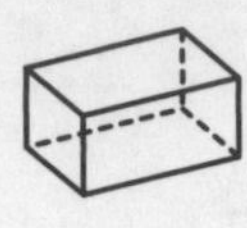  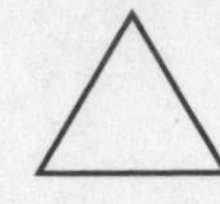 

A B C D

10. A doctor changed the dressings with bags of maggots over a regular period of time. What is the missing time?

    9:30 A.M., 11:00 A.M., ?, 2:00 P.M., 3:30 P.M.

    ____

11. The maggot at right has been transformed. What type of transformation is it?

    ____

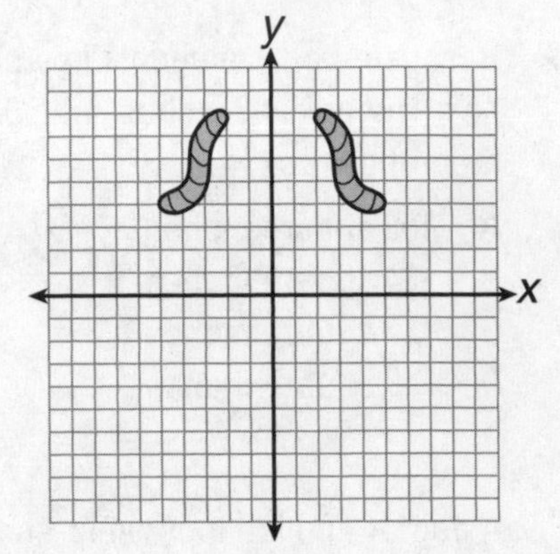

12. A maggot is squirming and its body is shaped at various angles. Which angle in the answer choices is congruent to the angle below?

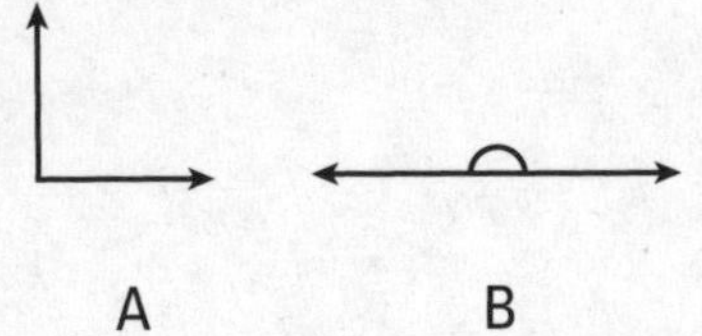 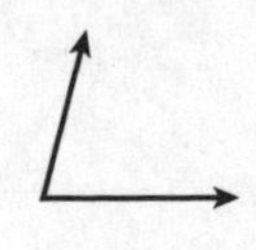 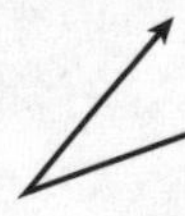

A B C D

CHAPTER 16

Measuring Objects and Converting Units of Measure

What comet can be seen approximately every 76 years?

What Major League Baseball prospect can throw a ball 168 miles per hour?

Units of Measure

There are many different tools to use for measuring. The ones you will use most often in school are rulers, protractors, and scales. **Rulers** measure the length of lines and objects. **Protractors** measure angles formed by two lines. **Scales** measure the weight of objects.

U.S. Customary System Units

Sometimes you'll need to convert measurements from one unit to another. Suppose you need to know the number of ounces in a pint. There are 2 cups in a pint, and 1 cup is 8 fluid ounces. $2 \times 8 = 16$, so there are 16 ounces in a pint. How many feet are in 4 yards? 3 feet = 1 yard, and $4 \times 3 = 12$, so there are 12 feet in 4 yards. How many square feet are in a square yard? A square yard has 3 feet on each side, so a square yard would be 3 feet × 3 feet = 9 square feet. There are 9 square feet in a square yard.

Length	12 inches (in.)	=	1 foot (ft.)
	3 feet (ft.)	=	1 yard (yd.)
	1 mile (mi.)	=	5,280 feet (ft.)
Weight/Mass	16 ounces (oz.)	=	1 pound (lb.)
	2,000 pounds (lb.)	=	1 ton
Liquid Volume	8 fluid ounces (fl. oz.)	=	1 cup
	2 cups	=	1 pint (pt.)
	2 pints (pt.)	=	1 quart (qt.)
	4 quarts (qt.)	=	1 gallon (gal.)

Metric System Units

You convert within the metric system the same way you do in the U.S. customary system. If you want to know how many millimeters are in a meter, you would multiply 10 millimeters times 100 centimeters to get 1,000 millimeters in a meter. How many centimeters are in 1.5 meters? $1.5 \times 100 = 150$. There are 150 centimeters in 1.5 meters.

Length	1 kilometer (kg)	=	1,000 meters (m)
	1 meter (m)	=	100 centimeters (cm)
	1 centimeter (cm)	=	10 millimeters (mm)
Weight/Mass	1 kilogram (kg)	=	1,000 grams (g)
Liquid Volume	1 liter (L)	=	1,000 milliliters (ml)

Comparing Units

You should be able to convert from one system of measurement to another. Unfortunately, it is not easy to convert the numbers, because the units are in such different systems. Converting from one system of measurement to another may be easier to remember if you know approximate comparisons. For instance, 1 quart is a little less than 1 liter, 1 yard is a little less than 1 meter, 1 mile is about 1.5 kilometers, and 1 kilogram is about 2 pounds.

When converting measurements, think about the unit you're converting to so that you'll know whether to multiply or divide. Remember that the number of inches will be greater than the number of feet units and that the number of meters will be less than the number of centimeters.

Directions: Read the passage below and answer the question that follows.

The Once in a Lifetime Comet

Termed "dirty snowballs" or "icy mud balls," comets are a mixture of rock, dust, ice, and gas that travel around outer space. Some comets travel egg-shaped orbits, which take them beyond Pluto. Others are seen once and never observed again. Some are seen from Earth on a regular basis, like Halley's comet, which was last seen in 1986 and will not be seen again until 2061. How old will you be the next time it comes around?

► Jenna and her parents were looking at a comet through the telescope below. What is the angle formed by the telescope and the ground underneath it?

____

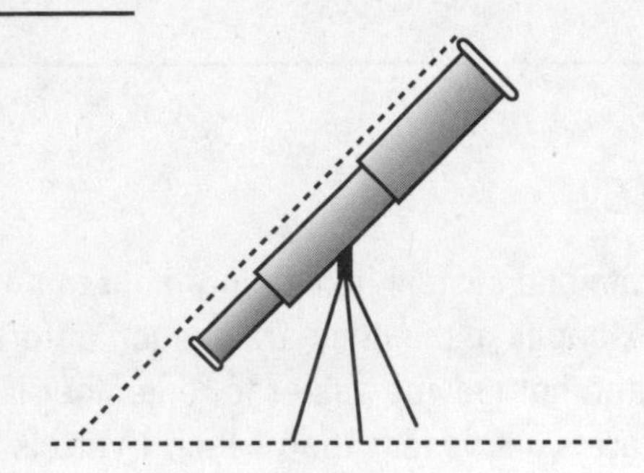

Know It All Approach

Read the question carefully and notice the information you'll need to get this one. The phrase "angle formed" is what you need to know. You will need to use a protractor to find the measurement of the angle.

Remember, the angle you are measuring is the angle between the ground and the arm of the telescope. You should find the angle to be 45 degrees. One way to double-check your answer is to draw your own 45-degree angle and compare it to the angle in the picture.

You may have measured the angle to be 135 degrees. This is the correct measurement for the angle formed by the telescope and the ground behind it. However, you're looking for the angle the telescope forms with the ground *underneath* it.

Write "45 degrees" clearly on the line.

Directions: Read the passage below and answer the questions that follow.

The Greatest Sports Hoax Ever

The April 1, 1985, edition of *Sports Illustrated* had an article about a rookie pitcher who planned to play for the New York Mets. His name was Sidd Finch, and he could accurately throw a baseball at 168 miles per hour, much faster than anyone else in history! The power of his pitches made them impossible to hit, and Mets fans all over celebrated. After the article was published, *Sports Illustrated* received thousands of letters requesting more information. Finally, the magazine admitted that the whole thing was a big April Fool's Day joke. The date the issue was published, after all, was April Fool's Day, 1985.

1. Sidd Finch's fastball was said to travel at 168 miles per hour. About how far did Sidd Finch's fastball travel in 1 second?

   A 36 feet
   B 52 feet
   C 246 feet
   D 887 feet

2. Use your ruler to find how far the baseball traveled in the diagram below.

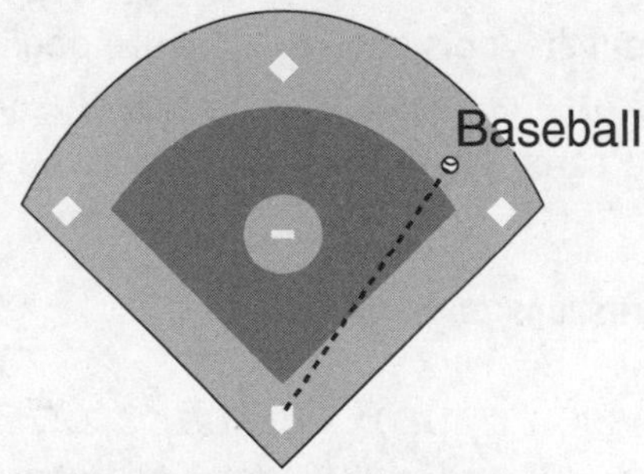

Scale: 1 inch = 30 yards

   A 40 yards
   B 60 yards
   C 30 yards
   D 20 yards

3. A trainer made some sports drink for his players. One container of concentrate holds 32 fluid ounces and needs to be mixed with 4 quarts of water. How many gallons of sports drink will one container make?

   A 1.25 gallons
   B 1 gallon
   C 12.5 gallons
   D 10 gallons

4. **Part A**

   A baseball is hit 156 feet. How many yards is 156 feet?

   ____

   Part B

   About how many meters is 156 feet?

   ____

5. If a particular baseball weighs 145 grams, how many kilograms will 20 such baseballs weigh?

   ____

Subject Review

In this chapter, you learned about the tools used to measure angles, length, and weight. You learned about units in the U.S. customary system as well as the metric system. You also practiced measuring with a protractor. Be careful when you measure to be exact as you can, especially on tests you may take.

Here are the answers to the questions on page 139.

What comet can be seen approximately every 76 years?

Halley's comet was seen in 1986 and in 1910. It will be seen again in 2061. Mark Twain, the famous author, was born and died in the year of Halley's comet.

What Major League Baseball prospect can throw a ball 168 miles per hour?

So far, no real player has been able to throw more than 103 miles per hour, but a fictional character named Sidd Finch learned to throw nearly 170 miles per hour in Nepal.

CHAPTER 17

Choosing Units of Measure and Estimating Measurement

How far would you have to walk to get to the pot of gold at the end of a rainbow?

What swimmer won 7 gold medals in the 1972 Summer Olympics?

Choosing Units of Measure

In the last chapter, you learned about the different measures for both metric and U.S. customary units. Now, you need to think about which unit of measurement is best for certain situations. To measure the fluid in a test tube, you'd use milliliters, not liters, because a test tube is much smaller than a liter. However, when you're talking about the amount of liquid in a punch bowl, you'd use liters, because a punch bowl is larger than a liter. Likewise, the length of a football field would be in meters, not kilometers, but the distance from Boston to San Francisco would be in kilometers, not meters.

When you want to determine which unit of measure to use, think about the unit that will give you the simplest measurement. For example, it is better to say that a house is 10 meters high rather than 10,000 millimeters.

When you calculate the area of a figure, you're multiplying a unit by a unit, which will give you **square units.** For instance, a rectangle that is 24 centimeters long by 10 centimeters wide will have an area of 240 square centimeters. When you calculate the volume of a figure, you multiply a unit by a unit by a unit, and you will receive an answer in **cubic units.** A rectangular prism that is 3 meters long by 2 meters high by 4 meters wide has a volume of 24 cubic meters.

Estimating Measurements

Sometimes it's easier to think of a common item when trying to gauge a measurement. A paper clip weighs about 1 gram, a doorway is about 2 meters high, a textbook weighs about 2 pounds or 1 kilogram, and it takes about 10 minutes to bake a batch of double-chocolate chewy-gooey cookies. A liter is about a quart, so it takes about 4 liters to make a gallon of milk (to go with the cookies). An inch is about 2.5 centimeters, which is how much someone's waistline may grow if they eat *all* of the cookies.

The measure of an angle can be estimated too, based on your knowledge of other angles. Look at the angles below.

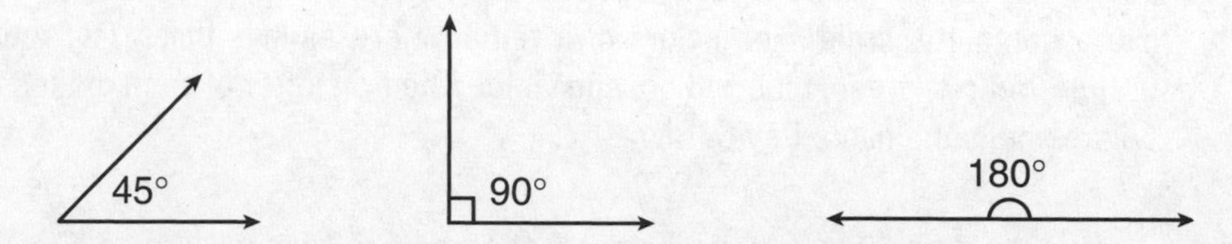

You can use the 45-degree angle as a basis for estimating the measure of angles from 0 degrees to 45 degrees and use the 90-degree and 180-degree angles for larger angles. If an angle is midway between 90 degrees and 180 degrees, you can estimate it at 135 degrees ($90^\circ + 45^\circ = 135^\circ$).

Directions: Read the passage below and answer the question that follows.

Who Is Roy G. Biv?

Do you know what causes a rainbow? A rainbow is what happens when light hits a drop of water. The water bends the light and, acting like a prism, breaks the light into different colors. Larger drops of water cause brighter rainbows. This is why you usually seen rainbows on a sunny day after a rain shower. The largest rainbows occur in the afternoon or morning, when the sun is closest to rising or setting. Did you know that from an airplane a rainbow doesn't always look like an arch? Often from such heights, a rainbow looks like a complete circle! The colors of a rainbow are always the same: red, orange, yellow, green, blue, indigo, and violet. The first letter of each of those colors spells the name Roy G. Biv.

► The pot of gold at the end of the rainbow belongs, according to Irish folklore, to a leprechaun. Leprechauns are grumpy Irish fairies that are said to be very short and resemble old men. What would be the best unit of measure for the height of a leprechaun?

A meters
B yards
C feet
D kilometers

Know It All Approach

Read the question carefully and look for the information you'll need to answer the question. The words to pay attention to are "very short" and "best unit of measure."

To calculate the answer, think about the height of the leprechaun. Since it is very short, the best units of measure are centimeters, inches, or maybe feet.

To double-check your answer, think about how many feet tall a leprechaun might be. Maybe 2 feet? Think about other objects you know of that are 2 feet and compare them in your mind to a leprechaun.

Read all of the answer choices and remember the list of units you wrote: centimeters, inches, feet. The only one that appears as an answer choice is feet. Answer choice (C) must be correct. You can eliminate answer choices (A) and (B) because they are almost the same measurement and are too large. You can eliminate answer choice (D) because it is way, way, way too big! Answer choice (C) is correct.

Directions: Read the passage below and answer the questions that follow.

Athletes are always making tremendous achievments, whether it's besting the competition in Grand Slam events in tennis or golf, hitting dozens of home runs in baseball, passing for thousands of yards in football, scoring numerous goals in soccer or hockey, or running a distance in blazing time in track. You'll see some of these amazing feats in the questions below.

1. The Dallas Cowboys played in 8 Super Bowls in the twentieth century, more than any other team, and they won 5 of them. All Super Bowl games have been played on football fields that are regulation size. If you were to measure the length of a football field, which of the following units would be best for you to use?

   A miles
   B inches
   C feet
   D centimeters

2. In 1996, Andrei Chemerkin won an Olympic weightlifting event called the clean-and-jerk. The weight he lifted is about the same as the weight of 4 adults. Which of the following is the best estimate of the weight that Andrei Chemerkin lifted in 1996?

   A 260 ounces
   B 260 kilograms
   C 260 tons
   D 260 grams

3. At the Olympic Games in 1972, Mark Spitz won 7 gold medals in swimming. About how much water would an Olympic pool probably hold?

   A 900,000 gallons
   B 90,000 milliliters
   C 90 gallons
   D 900 cups

4. In 1999, Maurice Green ran the 100-meter dash in 9.79 seconds. If a runner ran the length of a marathon, which of the following units would be best to use to measure the distance that he ran?

   A meters
   B kilometers
   C centimeters
   D millimeters

5. In 2002, Anni Friesinger from Japan skated 1,500 meters on ice in 1 minute, 54.02 seconds. Which of the following is the best estimate for the weight of a pair of ice skates?

   A 4 ounces
   B 4 grams
   C 4 pounds
   D 4 tons

6. In 1999, the United States women's soccer team won the World Cup, an international soccer competition. Which of the following units would be best to use to measure the weight of a standard soccer ball?

   A ounces
   B tons
   C pounds
   D meters

7. **Part A**

   A shot put is a very heavy metal ball that athletes throw over their shoulders. The athlete that throws the shot put the longest distance wins. What unit would be best used to weigh a shot put?

   ____

   Part B

   An athlete is throwing a shot put. Use your protractor to measure the angle of the throw.

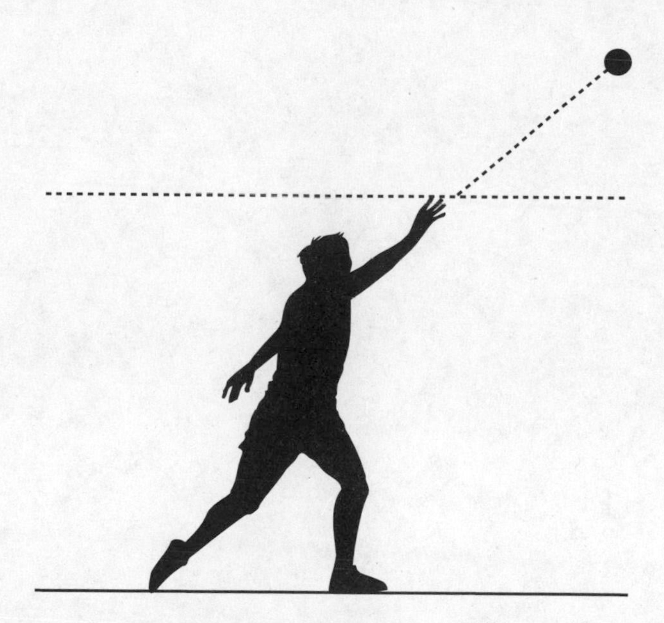

   ____

Subject Review

In this chapter, you learned about the units of measure and how to decide what unit is most appropriate for the type of measurement you are dealing with. You learned that when you multiply units together to get area or volume, the answer will be in square or cubic units. Measurements of area will be written in square units, and measurements of volume will be written in cubic units.

Remember the questions from page 145? Here are the answers.

How far would you have to walk to get to the pot of gold at the end of a rainbow?

A rainbow is made of light, so no matter how far you walked, you'd never find the pot of gold, because you can never get close to a rainbow.

What swimmer won 7 gold medals in the 1972 Summer Olympics?

In 1972, American Mark Spitz won 7 gold medals in swimming events at the Summer Olympics.

CHAPTER 18

Mean, Median, Mode, and Range

Which two players battled it out for the 1998 Major League Baseball home run title?

What was one of the most expensive natural disasters in history?

What type of storm can have winds stronger than 300 miles per hour?

Mean

The **mean** of a set of data is the sum of the numbers divided by the number of items of data. The mean of a set of data is the same as the data's average. The grades for a math test are 83, 68, 76, 80, 94, 76, and 90. There are 7 numbers in the set of data, so you would add the numbers and divide by 7 to get the mean. $\frac{83 + 68 + 76 + 80 + 94 + 76 + 90}{7} = \frac{567}{7} = 81$. The mean is 81.

Median

The **median** is the middle number in a set of numbers when the set is arranged in numerical order. Take the same set of numbers as above and put them in numerical order: 68, 76, 76, 80, 83, 90, 94. The median is the middle number, or 80. If the set of data consists of an even number of numbers, the median is the mean (average) of the two middle numbers. For instance, for the set 2, 4, 6, 9, the median is $\frac{4 + 6}{2} = 5$.

Mode

The **mode** is the number that appears most often in a set of numbers. To find the mode, just look at the set of numbers and see which one appears the most often. Take the math test scores again and go through them to see which number appears most often. It's 76, which is the only number that appears twice. Sometimes a set of data will have more than one mode, and sometimes it will have no mode.

Range

The **range** of a set of numbers is the difference between the greatest and least values. To find the range, place the numbers in numerical order. Since the math scores above were put in order to find the median, you can see that the greatest value is 94 and the least value is 68. Subtract the least from the greatest to find the range. $94 - 68 = 26$. The range is 26.

Finding the median and mode of a set of numbers is easy! No addition, subtraction, or any of that stuff. Just put the numbers in order from greatest to least. The middle number is the median. The number that appears most often is the mode.

Directions: Read the passage below and answer the questions that follow.

A Historic Chase

The sports world had never seen anything quite like the home run battle between Mark McGwire and Sammy Sosa in 1998. For 37 years, Roger Maris had held the record for the most home runs in a season when he hit 61 in 1961. But from the start of the 1998 season, it seemed that McGwire and Sosa were locked in. Each was determined to break the coveted record. By the end of June, about halfway through the season, McGwire and Sosa were about even after Sosa hit a record 20 home runs in the month. Each player seemed to have a good chance to challenge the record. The historic feat finally occurred on September 8, when McGwire belted his 62nd home run of the season. But just six days later, Sosa managed to tie him up, hitting his own 61st and 62nd home runs of the season on September 14. Now, the two had to battle each other for the new record with only two weeks left in the season. McGwire finally bested Sosa, finishing the season with 70 home runs, 4 more than Sosa's final total. However, baseball fans will not soon forget their historic race, and the fact that they had both broken a decades-old record within just a few days.

Part A

Mark McGwire hit very long home runs. The 5 longest he hit in 1998 were 545 feet, 527 feet, 511 feet, 509 feet, and 501 feet. What is the mean of these 5 home run distances?

____

Part B

What is the range of these 5 home run distances?

____

Know It All Approach

Read the question carefully and find the information you need to answer it. For Part A, the word to notice is "mean," and for Part B, the word to notice is "range."

There are two parts to the question, but both of them will require you to put the numbers in numerical order: 501, 509, 511, 527, 545.

To calculate your answer for Part A, you will need to find the mean of the 5 numbers. The mean is the average of all of the numbers. To find the mean, add the numbers together and then divide by the number of numbers there are, 5 in this case. $501 + 509 + 511 + 527 + 545 = 2{,}593$. If you divide 2,593 by 5, you will determine that the mean is 518.6.

In Part B, the range is found by subtracting the least number from the greatest number. $545 - 501 = 44$.

To double-check your answer, make sure that you have copied the set of numbers correctly and put them in the right order. On the lines after each part of the question, make sure that you write the numbers very clearly so that someone else can read them.

Directions: Read the passage below and answer the questions that follow.

Expensive Earthquake

Hurricanes, earthquakes, and tornadoes have always caused immense damage. Each of these destructive weather forces is measured with its own scale. One of the ways to record the damage level of a disaster is to calculate the amount of money it would cost to repair everything that was lost. One of the most expensive disasters in history occurred in Japan in 1995. It was an earthquake that was estimated to cost more than 131 billion dollars!

1. Earthquakes are measured on the Richter scale, which is a scale from 1 to 10 that measures the magnitude of an earthquake. The table below lists the readings for some of the most powerful earthquakes.

Date	Place	Richter Scale
1906	San Francisco	8.3
1920	China	8.6
1927	China	8.3
1952	California	7.7
1964	Alaska	8.5
1985	Mexico	8.1

Part A

What is the difference between the mean and median of the set of data about earthquakes?

____

Part B

What is the mode of the set of data?

____

2. The intensity of a hurricane can be rated on the Saffir-Simpson scale. A Category 1 storm has a wind speed of at least 74 miles per hour. The most devastating hurricane is a Category 5 storm, with winds over 155 miles per hour.

Date	Wind Speed
1935	200 mph
1969	160 mph
1987	115 mph
1988	175 mph
1992	150 mph

Part A

What is the mean wind speed of the hurricanes listed in the table above?

____

Part B

What is the range of the wind speeds of the hurricanes listed in the table above?

____

3. A storm surge, which is water pushed onshore, can be a damaging effect of a hurricane. If the storm surges of a group of hurricanes were 1.8 feet, 3.4 feet, 2.5 feet, 4.7 feet, 2.5 feet, 1.5 feet, and 3.2 feet, which of the following is the correct statement?

A mean = mode
B mean > mode
C range = mode
D range < mode

4. Tornadoes can be rated on the Fujita scale. The scale has six levels, from level F0, which has winds of 40 miles per hour, to level F5, which has winds up to 318 miles per hour. If the winds of five tornadoes are recorded at 230 miles per hour, 165 miles per hour, 195 miles per hour, 85 miles per hour, and 110 miles per hour, what is the mean wind speed?

____

5. Meteorologists recorded the following wind speeds from a series of tornadoes: 245, 55, 110, 95, 205, and 150 miles per hour. What is the median of these wind speeds?

A 135
B 120
C 130
D 125

Subject Review

In this chapter, you learned about the mean, median, mode, and range of a set of numbers. The mean is the average of a set of numbers. The median is the middle number of the set. The mode is the number that appears most often in the set. And the range is the difference between the greatest and least numbers in the set. You practiced finding the mean, median, mode, and range for a variety of sets of numbers.

And now, the long-awaited answers to the questions from the beginning of the chapter:

Which two players battled it out for the 1998 Major League Baseball home run title?

Mark McGwire and Sammy Sosa both broke Roger Maris's 37-year-old home run record in 1998. Both players hit their record-breaking home runs within a week of each other.

What was one of the most expensive natural disasters in history?

One of the most expensive disasters in history was a 1995 earthquake in Japan that was estimated to cost more than 131 billion dollars.

What type of storm can have winds stronger than 300 miles per hour?

Tornadoes have been known to have winds faster than 300 miles per hour.

CHAPTER 19

Graphs and Data

What dwarf is more than 200 billion years old?

Who holds the record for the most points scored in the National Basketball Association?

Bar Graphs

A **bar graph** uses bars to compare quantities. Suppose a survey conducted in your school determined what types of shows students watch on TV. You can use a bar graph to display the results.

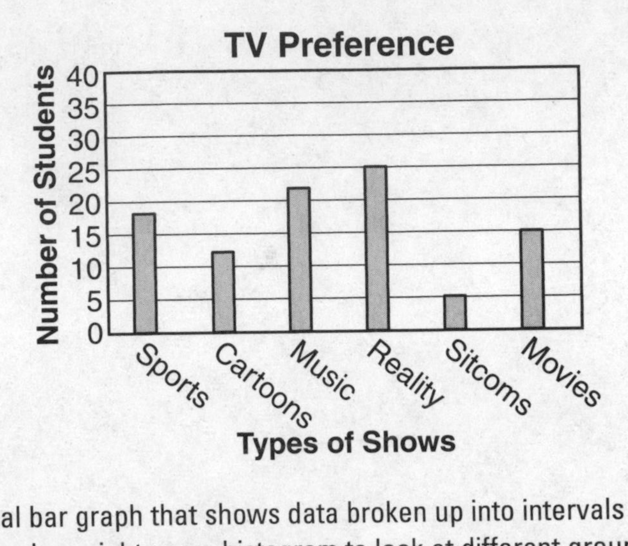

A **histogram** is a special bar graph that shows data broken up into intervals that cover the entire range of the data. A teacher might use a histogram to look at different groups of scores in the results of a speling—oops, that's *spelling*—test.

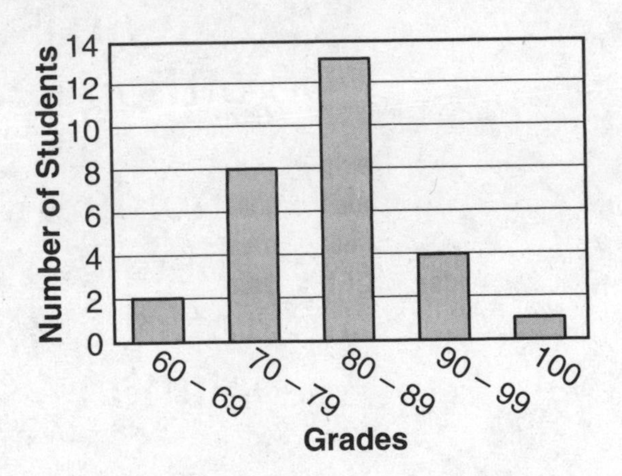

Reading a lot of numbers that tell you about a set of data usually doesn't help you to quickly understand what the information means. By placing the numbers in some type of graph, you get a much better picture.

Box Plots

A **box plot** compares data while showing numbers that are outside the boundaries of most of the set. Say the results of a spelling test were wide ranging, as follows: 86, 92, 33, 82, 68, 39, 64, 82, 68, 80, 78. The box plot of this data would be graphed as below. Note that the middle line of the box is at the median and that the lines from the box indicate the least and greatest test scores.

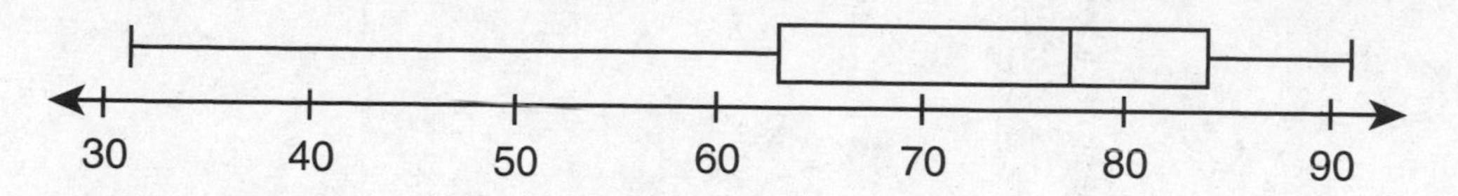

Scatterplots

A **scatterplot** shows the relationship between two sets of data. If one set of data increases as the other increases, then the relationship between the two sets of data is positive, and the graph will slope upward. The scatterplot in Figure A below shows the relationship between the height of a class of students and their arm span.

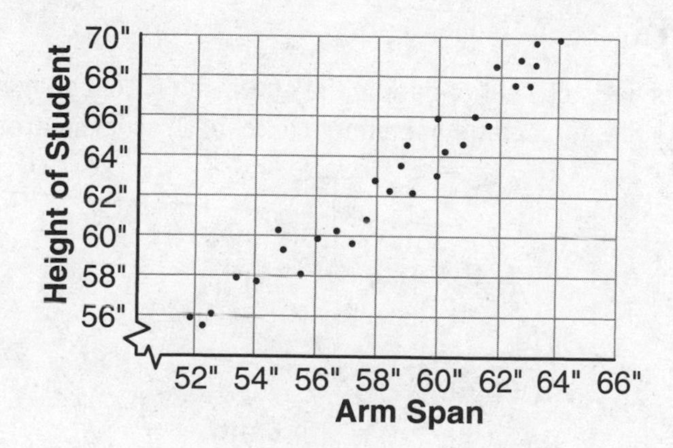

If one set of data increases while the other decreases, then the relationship of the data is negative, and the graph will slope downward. If student Thorton T. Throckleberry spends his time tapping on his tambourine instead of studying, Figure B, below, might be representative of his test scores.

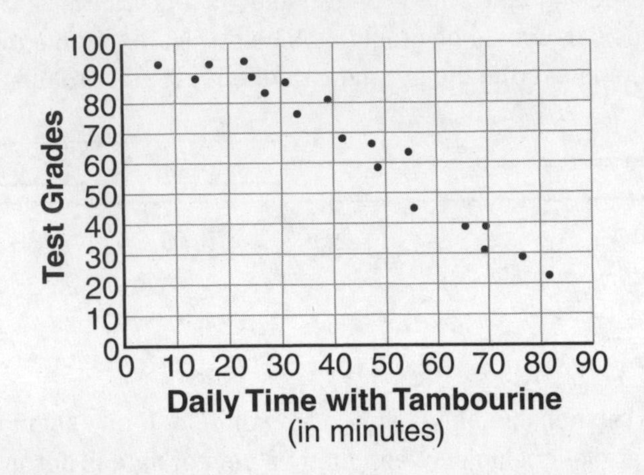

If a scatterplot has points that are widely scattered, there is no relationship between the sets of data.

Circle Graphs

A **circle graph** compares parts of a set of data to the entire set. A college may use a circle graph like the one below to see the percentage of students enrolled in six main areas of study.

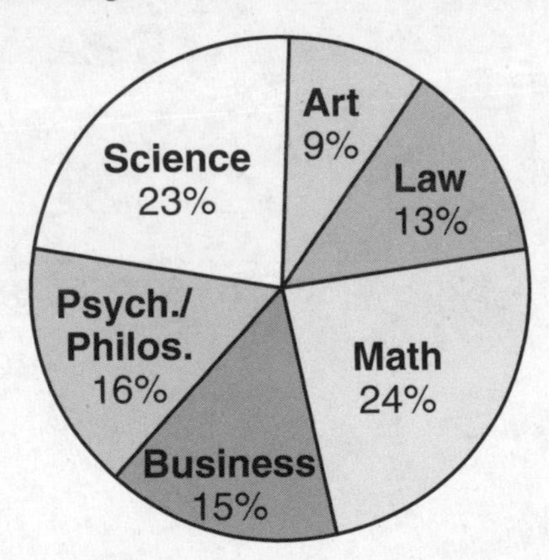

Directions: Read the passage below and answer the question that follows.

The Oldest Stars

Scientists use several different classifications for stars. One type is the dwarf star. This type of star lives the longest—200 billion years—but is often too faint to be seen because it produces very little energy. The oldest stars in the universe today are dwarf stars. An example of a dwarf star is Alpha Centauri, which is the closest star to Earth besides the Sun. Another type of star is a "normal" star, like the Sun. A normal star lives about 9 billion years. A third type of star is the giant star, which lives about 30 million years and produces a tremendous amount of light over a much shorter time span. Another type, the supergiant, has an even shorter life and produces even more light and energy. Rigel, in the Orion constellation, is an example of a supergiant.

▶ Look at the scatterplot below of some specific stars an astronomer is studying. Determine what the scatterplot tells you about the life of a star in relation to its mass.

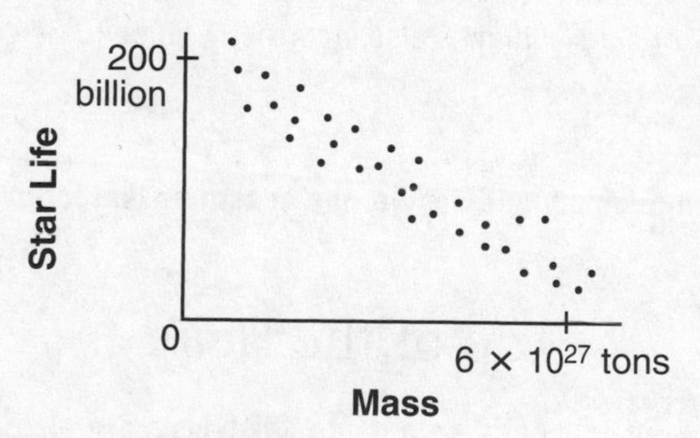

A as their mass increases, stars live longer
B mass doesn't affect the life of a star
C the stars that live the longest have the least mass
D all stars have the same mass

Know It All Approach

Read the question carefully and figure out the information you'll need. The words "the life of a star in relation to its mass" is the necessary information here.

To calculate the answer, look at the numbers on the axes of the scatterplot. You can see that the years increase to more than 200 billion years and the mass increases to more than 6×10^{27} tons.

Now, look at how the plotted points relate to those two scales. There's a negative relationship between them. What does that mean? The oldest stars have the least mass, and the stars that die young (30 million years is young?) have the greatest mass.

To double-check your answer, check to see that you have interpreted the information correctly. Is a lightweight star older? Is the star with the most mass the quickest to die?

Read all of the answer choices and use Process of Elimination. Answer choice (A) says that the more massive stars live longer. That's not right. Eliminate it. Answer choice (B) says that the mass of a star has no relation to its life. You can see that isn't true because the scatterplot "line" does change. Eliminate it. Answer choice (C) says that the oldest stars have the least mass, which is true. That's the correct answer! But check answer choice (D). It says that there is no relationship between the life and mass of stars, which is also incorrect. Choice (C) is correct.

Directions: Read the passage below and answer the questions that follow.

Top of the Heap

Pretty much anything can be put into a list. Lists are also a great source of data for mathematical calculations and charts. The following questions all relate to various lists. You will see that data related to science, sports, and art can all be displayed in a list.

1 **Part A**

The top five highest-priced paintings of all time, as of 2001, were painted by four artists: Paul Cézanne, Pablo Picasso, Pierre-Auguste Renoir, and Vincent van Gogh. A stationery store sells postcards of the famous painters' work. The circle graph below shows the percentage of postcards sold of each painter's work. Which painter's postcards are the best selling?

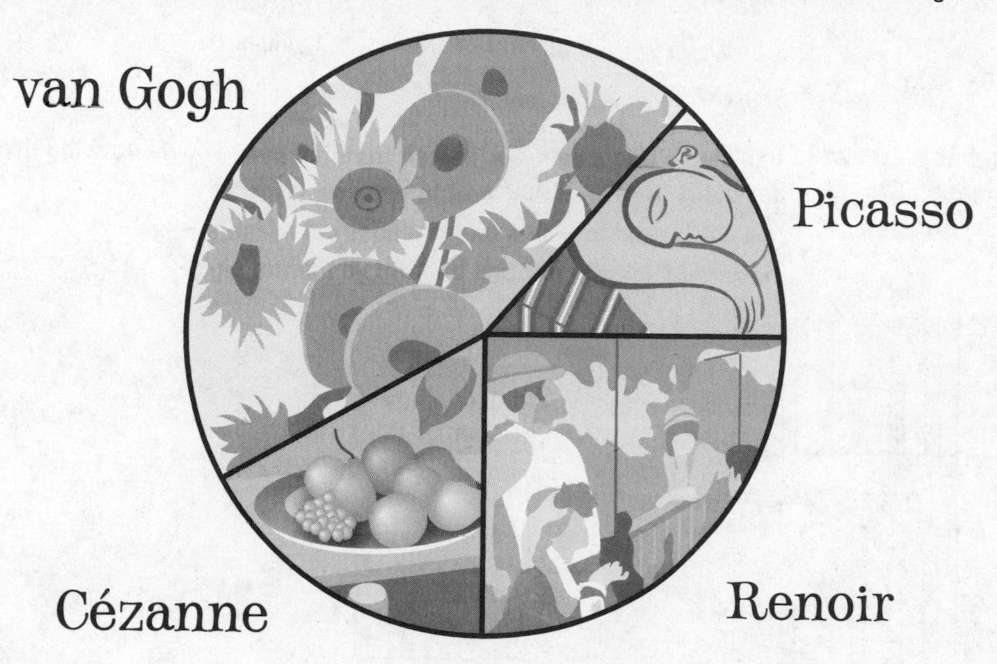

____

Part B

How do you know which painter's postcards sell the most copies?

____

____

____

2. As of 2002, the top money-making movies of all time in the U.S., in order of release dates, were as follows:

Star Wars	$461 million
E.T.	$435 million
Titanic	$601 million
Star Wars: The Phantom Menace	$432 million
Spider-Man	$404 million

If a bar graph were used to display the data, which of the following shows how the graph would look? (The order of the bars matches the order above.)

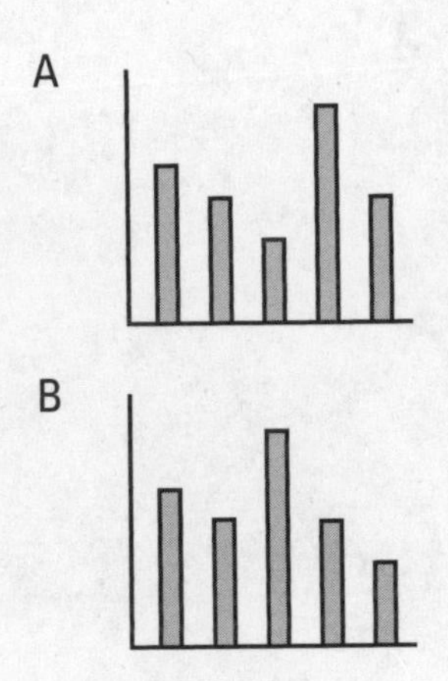

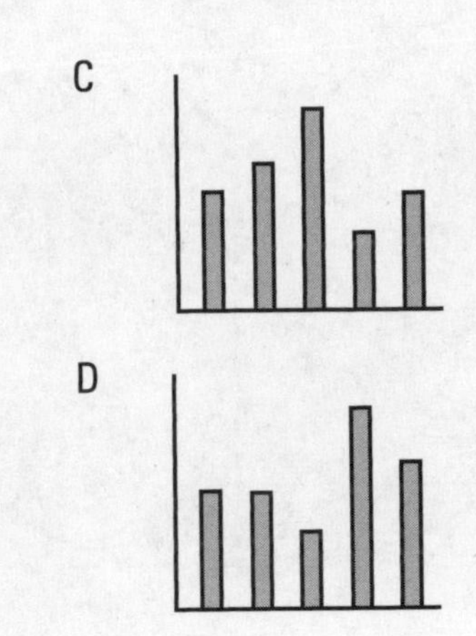

3. Which graph would be the best to display the wide range of ages of people who attended the first showing of Star Wars at a local movie theater?

A scatterplot
B a bar graph
C a box plot
D a circle graph

4. The Los Angeles Lakers have been one of the top teams in the history of the National Basketball Association. Between 2000 and 2002, the Lakers won three consecutive NBA championships. The 2003 team players are listed below with their heights and average of points per game in 2003.

Kobe Bryant	6′6″	30.0	Tracy Murray	6′7″	2.0
Derek Fisher	6′1″	10.5	Shaquille O′Neal	7′1″	27.5
Rick Fox	6′7″	9.0	Jannero Pargo	6′1″	2.5
Devean George	6′8″	6.9	Kareem Rush	6′6″	3.0
Robert Horry	6′10″	6.5	Brian Shaw	6′6″	3.5
Mark Madsen	6′9″	3.2	Samaki Walker	6′9″	4.4
S. Medvedenko	6′10″	4.4			

Part A

Would a scatterplot with the relationship between the heights and scoring record be positive, be negative, or have no relationship? Why?

____

____

____

Part B

On a separate sheet of paper, draw a histogram of the team heights with the intervals 6′0″ to 6′3″, 6′4″ to 6′7″, 6′8″ to 6′11″, and 7′0″ to 7′3″.

5. At the time of Michael Jordan's retirement from the NBA at the end of the 2003 season, only two players in history had scored more points in their career than him. The bar graph below shows the top five scorers in the NBA as of the 2003 season. Which of the players has scored the **least** number of points of the top five?

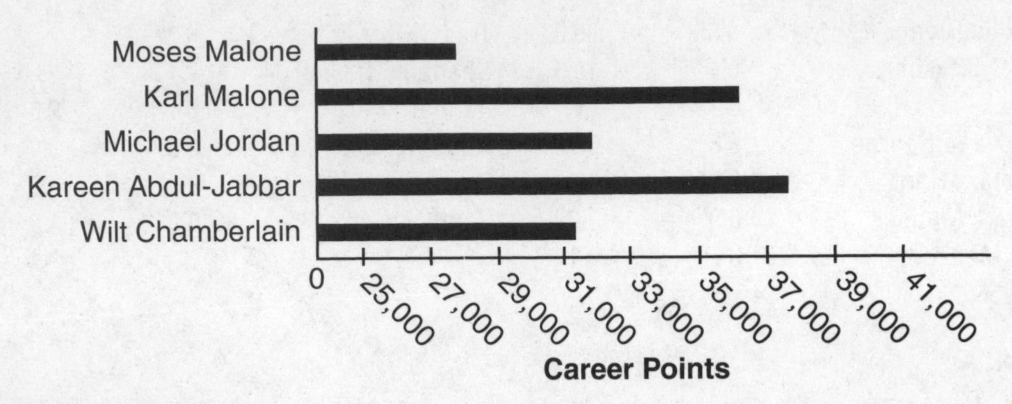

A Moses Malone
B Wilt Chamberlain
C Karl Malone
D Michael Jordan

Subject Review

In this chapter, you learned about several different types of graphs. Bar graphs and histograms are used to compare different categories of data. Box plots summarize a set of data, indicating the median and the extremes of the greatest and least numbers. Scatterplots are useful for comparing two characteristics of a single population. Circle graphs compare parts of a set of data to the whole.

Here are the answers to the questions from page 159.

What dwarf is more than 200 billion years old?

Dwarf stars are the oldest known stars in the universe. They will exist for approximately 200 billion years, and the universe is only 12 to 14 billion years old now!

Who holds the record for the most points scored in the National Basketball Association?

Kareem Abdul-Jabbar's record of 38,387 points is the most scored in the history of the NBA.

Probability

How many people did it take to make a bucket brigade 11,471 feet long?

Where can you get a scoop of octopus ice cream?

Why do male Egyptian vultures eat poop?

Simple Events

Probability is the chance that a certain event will happen. Probability can be expressed by using a ratio. Suppose there is a penny, a nickel, a dime, and a quarter in a cup. What is the probability that you would randomly choose the dime? There is 1 possible way the event can happen, because there is 1 dime. Because there are 4 coins, there are 4 possible events. The probability of choosing a dime, therefore, is 1 out of 4. This ratio can also be expressed as a fraction $\left(\frac{1}{4}\right)$, a decimal (0.25), or a percent (25%). Probability is the number of ways an event can happen in relation to the number of possible outcomes.

Simple and Compound Events

A **simple event** is an event that has a single set of outcomes. A **compound event** is an event that has more than one set of outcomes. An example of a compound event is the tossing of a penny and a nickel into the air. What is the probability that both coins will land heads up? The probability that the penny will land on heads is $\frac{1}{2}$. There is 1 favorable outcome (heads) to 2 possible outcomes (heads or tails). The probability that the nickel will land heads up is also $\frac{1}{2}$. Because the results of the first event, tossing the penny, don't affect the results of the other event, tossing the nickel, these two events are **independent events.** To find the probability of two independent events, multiply the probability of the first by the probability of the second: $\frac{1}{2} \times \frac{1}{2} = \frac{1}{4}$. The probability that both the penny and the nickel will land heads up is $\frac{1}{4}$.

One way to see all of the possible outcomes of a compound event is with a tree diagram such as the one below.

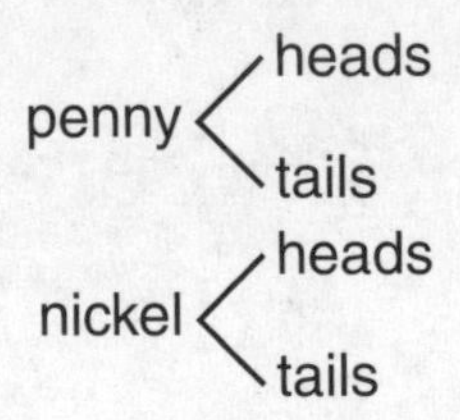

Compound events can also be dependent. For **dependent events,** the outcome of one event affects the outcome of the other. For example, suppose that 2 pennies, 2 nickels, and 2 dimes are placed in a cup. What is the probability that the one coin you pick is a nickel? There are 2 nickels out of 6 coins, so the probability is $\frac{2}{6}$.

Now, you pick another coin. What's the probability that the second coin is a nickel? There's 1 nickel left out of 5 coins left, so the probability is $\frac{1}{5}$.

To find the probability of both events occurring, multiply the probability of the first by the probability of the second. The probability of choosing two nickels is $\frac{2}{6} \times \frac{1}{5} = \frac{2}{30} = \frac{1}{15}$.

The tree diagram below shows all of the possible outcomes of picking two coins from the cup.

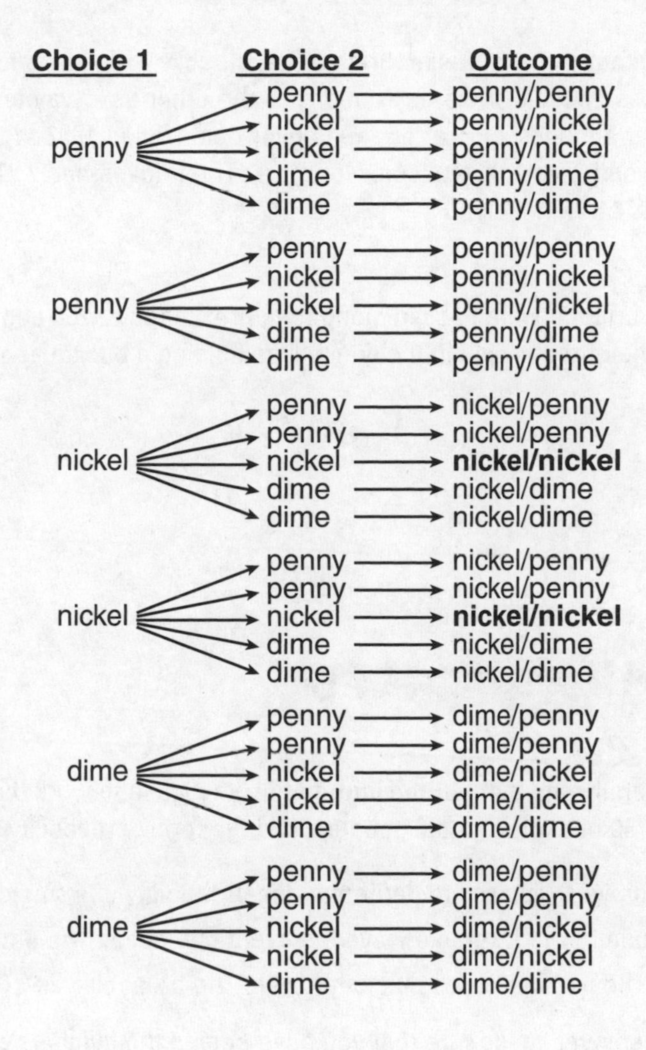

Directions: Read the passage below and answer the question that follows.

Call In the Brigade!

Do you know what a bucket brigade is? A bucket brigade is a line of people who pass buckets of water from one to the other as a way to put out a fire. One of the longest bucket brigades ever occurred in 1992 at the Centennial Parade and Muster at Hudson, New York. The brigade had 2,271 participants and was 11,471 feet long.

▶ 50 buckets in the brigade were passed along to approximately 2,200 people. What is the probability that one of the people in the line will be holding a bucket at a specific moment?

A 440 to 1

B $\frac{1}{44}$

C 44 to 1

D $\frac{1}{440}$

Know It All Approach

Read the question carefully and pick out the information you'll use to work this one out. The information to note is "50 buckets," "2,200 people," and the term "probability."

To calculate the answer, you will need to determine the probability of a single event. There are 50 buckets, which is the number of favorable ways the event can occur. The other part of the ratio is 2,200 people, which is the number of possible outcomes. The probability is $\frac{50}{2,200} = \frac{5}{220} = \frac{1}{44}$.

To double-check your answer, make sure that you have copied the numbers correctly and that you have put them into the ratio in the proper order. By reading all of the answer choices, you can determine that one of the answer choices is the same as the answer you got. Excellent!

However, use Process of Elimination because it is a good idea to look at the answer choices and decide why they're wrong (and make sure yours is right). You can eliminate answer choices (A) and (C) because they have the ratio in the wrong order. You can eliminate answer choice (D) because it is an answer that was not correctly simplified. Answer choice (B) is correct.

Directions: Read the passage below and answer the questions that follow.

A reality TV show tossed worms and roaches into a blender to make a lip-smacking shake. Don't you hope that McDonald's offers that soon? Another delicious serving was a plateful of the eyes of fish, sheep, and cows. Can you imagine how delicious a treat that must have been? The following questions all relate to foods that might seem more than a little bit gross.

1. **Part A**

In some parts of the United States, ice cream is served in flavors that might seem a bit, well, vile. Some such flavors include garlic, minestrone, sauerkraut, and praline chili. If scoops of each flavor above are placed in separate cups, what is the probability that you will randomly choose the cup with garlic ice cream?

____

Part B

In Japan, ice cream comes in some fabulous fishy flavors like shark fin, octopus, crab, salmon, and sea urchin. If you put scoops of these flavors into separate cups and place them alongside the American flavors mentioned above, what is the probability that the one cup you select will be from Japan?

____

2. Female Egyptian vultures seem to be attracted to male vultures who have faces that are bright yellow. The way a male vulture develops a bright yellow face is by eating cow, goat, and sheep poop. The more poop the vulture eats, the more yellow its face gets. It kind of makes you wonder what their breath smells like! If there are 6 male vultures with bright yellow heads and they are named Squidbreath, Poopface, Bigbeak, Smelly, Butthead, and Yucky, what's the probability that a female will pick a male vulture whose name starts with S?

   A $\frac{1}{6}$
   B $\frac{2}{5}$
   C $\frac{1}{2}$
   D $\frac{1}{3}$

3. In a jar, there are a bunch of tasty treats, and you have to eat one chosen at random. There are 15 dog biscuits, 15 hard-boiled crocodile eggs, and 20 maggots. What is the probability that a maggot is selected as the treat that you have to eat?

   A 50%
   B 35%
   C 40%
   D 20%

4. Suppose that you are required to eat a second treat from the jar. What is the probability that the second treat you have to eat is also a maggot?

   A $\frac{19}{49}$
   B $\frac{20}{49}$
   C $\frac{19}{50}$
   D $\frac{20}{50}$

5. Suppose there is a bowl with 27 cockroaches inside. The cockroaches are numbered from 1 to 27 on their backs. Now, you have to reach in and pick up a cockroach. What is the probability that you'll pick one with a number that's a multiple of 3?

____

6. Edwin wants to go out to buy some chicken livers, goose gizzards, and turkey wattles. He goes to the store with a couple of dollar bills and 3 quarters. If Edwin tosses all 3 quarters into the air, what is the probability that they will all come up heads?

A $\frac{1}{6}$

B $\frac{1}{4}$

C $\frac{1}{8}$

D $\frac{1}{3}$

7. A bowl in Edwin's kitchen contains 3 chicken livers, 3 goose gizzards, and 3 turkey wattles. Edwin's sister, Willow, has to pick an object from the bowl and hand it to Edwin, who will eat it. A goose gizzard was randomly chosen by Willow, and Edwin ate it. What is the probability that on the second pick Willow will choose another goose gizzard?

____

Subject Review

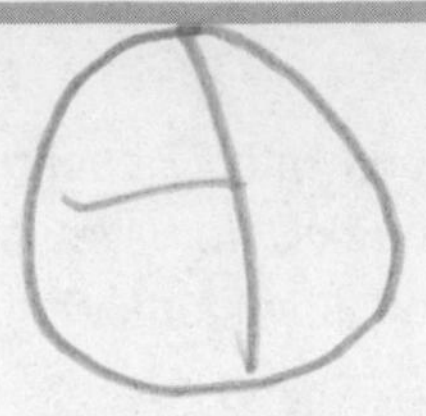

In this chapter, you learned about probability, but wasn't it fun? Lots of gross stuff to eat and touch! Remember, a simple event is an event that has only one set of outcomes. The probability of a simple event is found by putting the number of ways an event can occur over the number of possible outcomes. The result can be expressed as a ratio, a decimal, a fraction, or a percent. A compound event is an event made up of more than one simple event. The probability of a compound event is found by multiplying the probability of the first event by the probability of the second event. A dependent event is an event in which the outcome depends on the outcome of another event. The probability of two dependent events is found by multiplying the probability of the first event by the probability of the second event after the first event has occurred.

Without further ado, here are the answers to the questions on page 169.

How many people did it take to make a bucket brigade 11,471 feet long?

The bucket brigade created in Hudson, New York, in 1992 was 11,471 feet long and contained 2,271 people.

Where can you get a scoop of octopus ice cream?

In Japan, there are many popular seafood-flavored ice creams. Can you imagine ordering an octopus and squid sundae?

Why do male Egyptian vultures eat poop?

When male Egyptian vultures eat poop from cows, goats, and sheep, it turns the skin on their faces yellow, which is very attractive to female Egyptian vultures. Go figure!

Directions: Read the passage below and answer the questions that follow.

Not-So-Itsy-Bitsy Spiders

Tarantulas are big, hairy spiders. They can be found in many places but are especially prolific in the Amazon basin, where many species of them are still unknown to science. They cannot eat solid food, so they pump venom into their prey with their fangs. Venom is like a digestive fluid that turns flesh into liquid. Then, the tarantula sucks up the liquid. Tarantulas are able to kill everything from cockroaches to frogs to bats, but out of about 1,500 species, only 25 would be poisonous enough to kill a person.

1. Tarantulas can live for more than 20 years, and some can grow as large as a dinner plate. Which unit would you use to measure the size of a tarantula?

   A miles
   B millimeters
   C feet
   D centimeters

2. The National Reptile Breeders' Expo sells all sizes of tarantulas each year. Some spiderlings (baby spiders) can sell for $150. People like to keep them for pets. Lucy has the following types of pet tarantulas up for sale: 22 Ecuadorian purple, 26 Mexican blond, 18 South American goliath birdeater, 22 Peruvian pinktoe, 14 Chilean rose hair, 6 Mexican redknee, and 4 Mexican cave tarantulas.

   Part A

   What is the mode of Lucy's pet tarantulas?

   ____

   Part B

   What is the median of Lucy's pet tarantulas?

   ____

3. There are many kinds of tarantulas. Of those listed below, what is the probability that a tarantula chosen at random will start with the letter M?

Peruvian pinktoe
Mexican blond
Ecuadorian purple
Mexican cave
Chilean rose hair
South American goliath birdeater
Mexican redknee

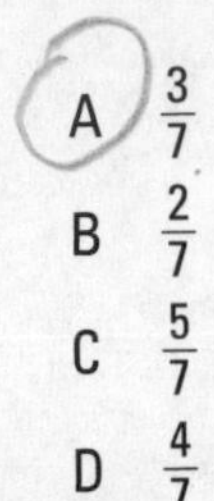

A $\frac{3}{7}$

B $\frac{2}{7}$

C $\frac{5}{7}$

D $\frac{4}{7}$

4. In a zoo, there's a collection of three kinds of tarantulas: African redrump, Javan yellowknee, and American curlyhair.

African redrump is $\frac{3}{6}$ of the collection.

Javan yellowknee is $\frac{1}{4}$ of the collection.

American curlyhair is $\frac{2}{8}$ of the collection.

Draw a circle graph using the information about the zoo's collection. Indicate which part of the graph represents which kind of tarantula.

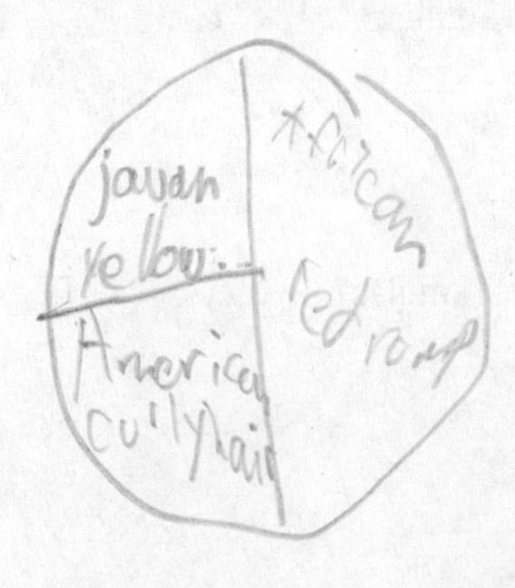

5. Pete's pet tarantula, Gargantua, needs more soil in his tank. Pete added 1 pint of soil to Gargantua's house. Which of the following measurements is equivalent to 1 pint?

A 2 ounces
B 2 cups
C 2 liters
D 2 gallons

6. Tarantulas have external skeletons, which they must shed so they can grow. If Eleanor's Mombasa golden starburst sheds its skeleton according to the table below, what is the mean growth per shed?

Times Shed	1	2	3	4	5	6	7	8	9
Growth	1.5 in.	1.2 in.	1.4 in.	0.8 in.	1.1 in.	0.5 in.	1.3 in.	1.1 in.	1 in.

A 1 in.
B 0.9 in.
C 1.1 in.
D 1.2 in.

7. **Part A**

Tarantulas kept as pets can be fed crickets. If 5 crickets called Moe, Larry, Curly, Chirpy, and Loudmouth are placed in a tarantula's container, what is the probability that the tarantula will eat Chirpy first?

____

Part B

What is the probability that the tarantula will eat Chirpy and then eat Moe?

____

8. A tarantula doesn't weigh a whole lot. What measurement would be most appropriate for the weight of a tarantula?

A pounds
B kilograms
C ounces
D tons

9. A pet supplier keeps a graph of the 4 types of tarantulas he sells each year. If 248 tarantulas were sold the entire year, how many were king baboon?

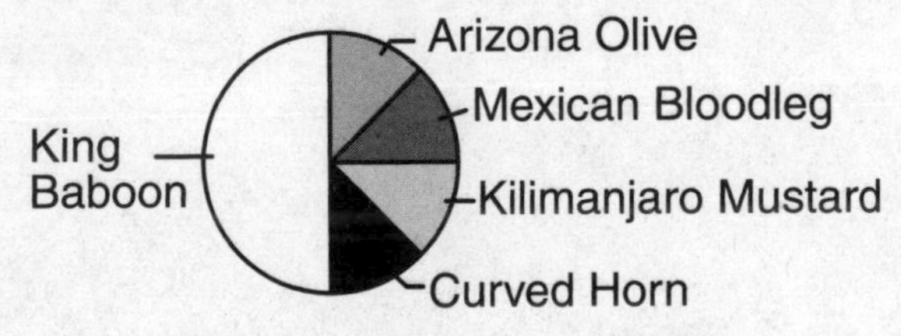

Percent of Tarantulas Sold

A 62
B 124
C 100
D 31

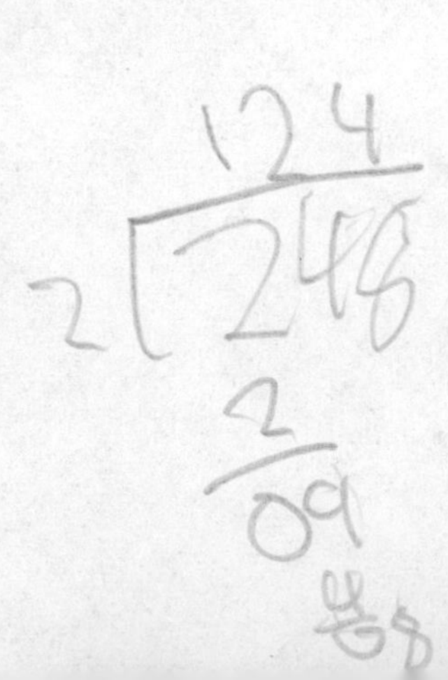

Answer Key for Chapters and Brain Boosters

Answer Key for Chapters and Brain Boosters

Chapter 2

1. D
2. B
3. 2.8 tons per foot
4. A. $35.\overline{185}$
   B. rational
5. B
6. D
7. 6
8. A. $.4\overline{09}$ seconds
   B. rational
9. The pangolin's body is approximately 3 times longer than its tongue.

Chapter 3

1. D
2. A. the car
   B. the motorcycle's
3. D
4. B

Chapter 4

1. B
2. A
3. 1,875 pounds
4. B
5. 9 sandwiches
6. C

Chapter 5

1. D
2. 3 to 8
3. C
4. B
5. 500 meters
6. 300 meters
7. A

Chapter 6

1. C
2. 2, 4, 7, and 14
3. D
4. A
5. 2, 5, 10, and 25
6. C

Chapter 7

1. C
2. 250
3. 49
4. B
5. D

Chapter 8

1. C
2. B
3. The following is one possible answer. The following numbers added in any order will produce a correct answer: $(1.20 + 0.75 + 0.35) + (1.20 + 0.90 + 0.25) = (1.20 + 0.75) + (0.35 + 1.20) + (0.90 + 0.25)$

Brain Booster 1

1. C
2. 33 million copies
3. D
4. B
5. 2 to 3
6. $6(16.99) + 3(16.99)$
7. A
8. C
9. 15 to 2
10. 12
11. D

Chapter 9

1. 25
2. C
3. $360 \div 8 = n - 2$

   The number of snakes killed by the explorers is 33.
4. B
5. D
6. A

Chapter 10

1. B
2. 17
3. 92
4. 450 liters
5. Acme Water Company
6. The pattern is that the waves increase $\frac{1}{2}$ foot every 3 minutes.

Chapter 11

1. B
2. C
3. To answer this question correctly, you should draw a rectangle with the word *rectangle* written underneath.
4. B
5. C
6. D
7. triangle
8. 6 sides
9. To answer this question correctly, you should draw a cone with the word *cone* written underneath.
10. 6
11. 8
12. C

Chapter 12

1. A. 792 square inches
   B. 116 inches
2. C
3. D
4. 84 square inches
5. A
6. B
7. A. 81
   B. 282

Chapter 13

1. B
2. To find the volume of the coffin, multiply the length (2 meters) by the width (1.5 meters) by the height (1 meter) of the coffin.
3. 96 blocks
4. D
5. C
6. A. 4 times (1,026 square inches)
   B. 8 times (2,268 cubic inches)
7. C
8. No, the difference is about 13 square inches.

Chapter 14

1. C
2. Congruent means that figures are exactly the same size and shape. The two shields are not congruent, because they are not the same size.
3. On graph paper, you should have drawn a rectangle that is 8 squares by 16 squares.
4. B
5. D

Chapter 15

1. A. The letter L is reflected.
   B. The transformed image is a reflection because it is a mirror image. The L was flipped.
2. A. Example of a correct rotation:

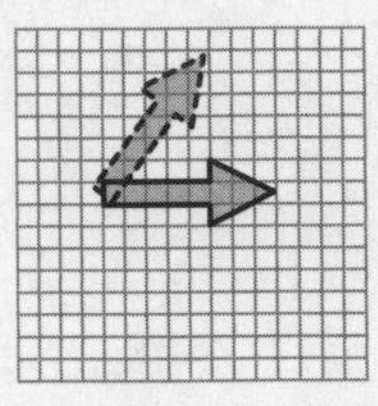

   B. Your answer is correct if your arrow base remains nearly in the same position but the arrow points in a different direction.
3. A
4. The easel has been reflected or flipped over the y-axis.
5. C

Brain Booster 2

1. C
2. 625
3. B
4. A. 48.96 square centimeters
   B. 28 centimeters
5. A. $50 + 20x = 750$
   B. 35
6. D
7. A
8. B
9. C
10. 12:30 P.M.
11. The transformation is a reflection.
12. D

Chapter 16

1. C
2. C
3. A
4. A. 156 feet is 52 yards.
   B. 156 feet is about 50 meters.
5. 2.9 kilograms

Chapter 17

1. C
2. B
3. A
4. B
5. C
6. C
7. A. pounds
   B. 40°

Chapter 18

1. A. 0.05
   B. 8.3
2. A. 160 miles per hour
   B. 85
3. B
4. 157 miles per hour
5. C

Chapter 19

1. A. van Gogh

   B. van Gogh's section is almost half of the circle, which is much larger than any of the other sections.
2. B
3. C
4. A. There is no relationship, because someone who is 6'1" may score higher or lower than someone who is 6'7" or 6'9".

   B. Your histogram should look similar to the following:

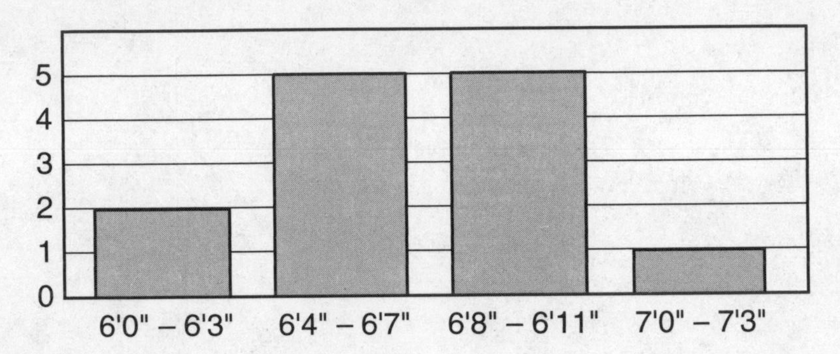

5. A

Chapter 20

1. A. $\frac{1}{4}$

   B. $\frac{5}{9}$
2. D
3. C
4. A
5. $\frac{1}{3}$
6. C
7. $\frac{1}{4}$

Brain Booster 3

1. D
2. A. 22
   B. 16
3. A
4.

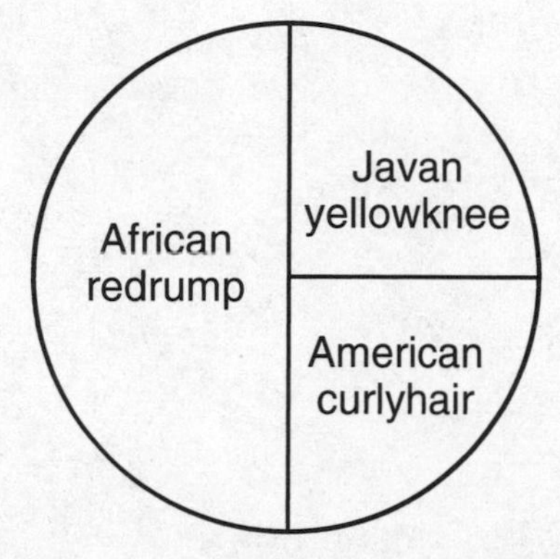

5. B
6. C
7. A. $\frac{1}{5}$
   B. $\frac{1}{9}$
8. C
9. B

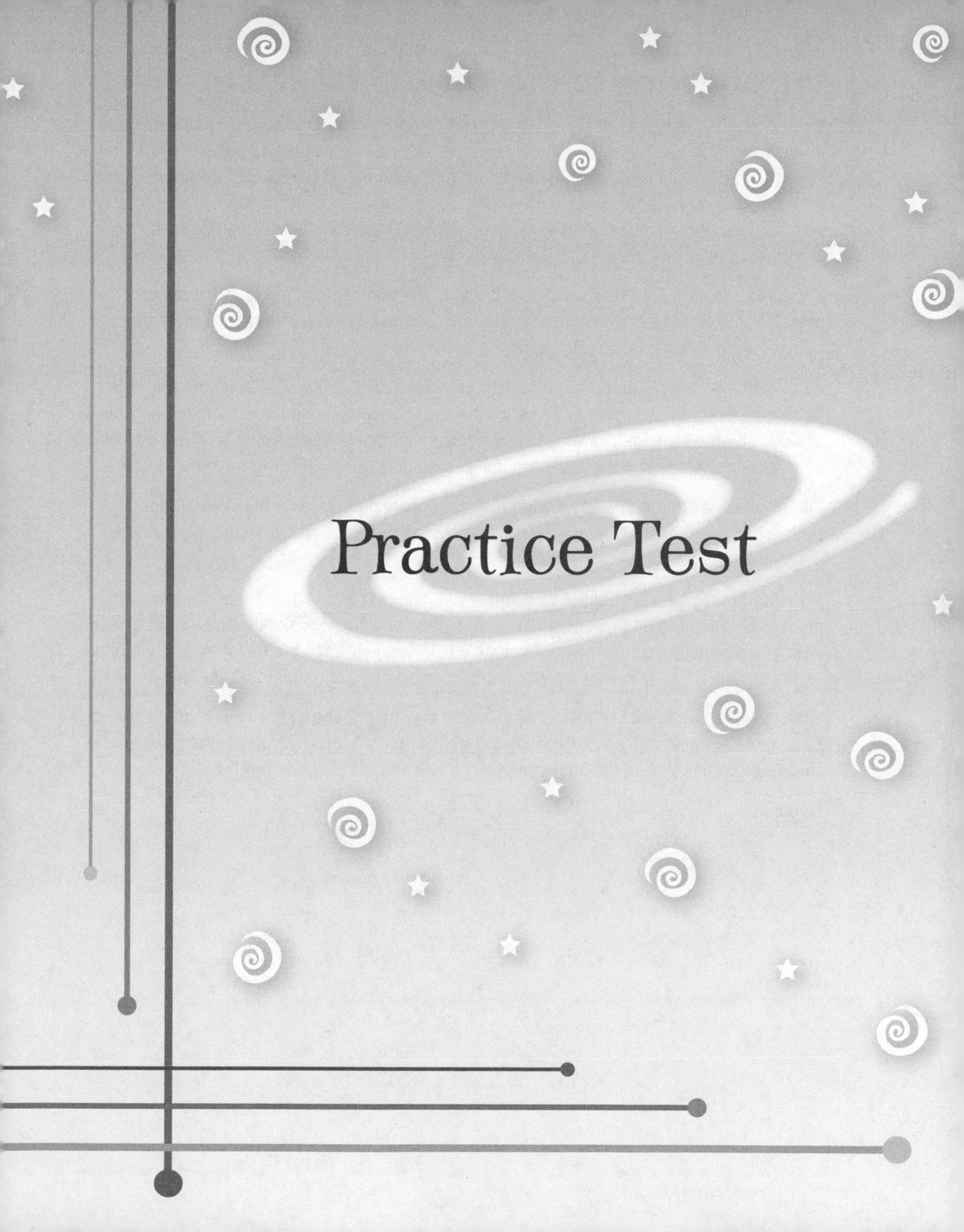

Practice Test

Introduction to the *Know It All!* Practice Test

By now you've reviewed all the important skills that you should know for middle school math. You know how to convert and compare fractions, decimals, and percents (chapter 3). You know how to extend patterns and find rules for functions (chapter 9). You also know how to find the area, perimeter, surface area, and volume of various figures (chapters 12 and 13). And these are just a few examples that don't even include all the excellent tidbits of information you've picked up. *You know it all*!

If you're ready, it's time to try out the skills from the twenty chapters in this book in a practice test. This test may be similar to a test you take in class. It contains multiple-choice, short-answer, and open-response questions.

Each multiple-choice question on the test has four answer choices. You should fill in the bubble for the correct answer choice on the separate answer sheet. Cut or tear out the answer sheet on the next page, and use it for the multiple-choice questions. You can write your answers to the short-answer and open-response questions directly in the test.

The practice test contains forty-two questions, including thirty-two multiple-choice questions, five short-answer questions, and five open-response questions. Give yourself ninety minutes to complete the test.

Take the practice test the same way you would take a real test. Don't watch television, don't talk on the telephone, and don't listen to music while you take the test. Sit at a desk with a few pencils, and have an adult time you if possible. Take the test in one day and all in one sitting. If you break up the test in parts, you won't get a real test-taking experience.

When you've completed the practice test, you may go to page 217 to check your answers. Each question also has an explanation to help you understand how to answer it correctly. Don't look at this part of the book until you've finished the practice test!

Good luck!

Bubble Sheet

1. use line provided
2. (A) (B) (C) (D)
3. (A) (B) (C) (D)
4. use line provided
5. (A) (B) (C) (D)
6. use line provided
7. (A) (B) (C) (D)
8. (A) (B) (C) (D)
9. (A) (B) (C) (D)
10. use line provided
11. use line provided
12. (A) (B) (C) (D)
13. (A) (B) (C) (D)
14. (A) (B) (C) (D)
15. (A) (B) (C) (D)
16. (A) (B) (C) (D)
17. (A) (B) (C) (D)
18. (A) (B) (C) (D)
19. (A) (B) (C) (D)
20. (A) (B) (C) (D)
21. use line provided
22. use line provided
23. (A) (B) (C) (D)
24. (A) (B) (C) (D)
25. (A) (B) (C) (D)
26. (A) (B) (C) (D)
27. use line provided
28. (A) (B) (C) (D)
29. use line provided
30. (A) (B) (C) (D)
31. (A) (B) (C) (D)
32. (A) (B) (C) (D)
33. (A) (B) (C) (D)
34. (A) (B) (C) (D)
35. (A) (B) (C) (D)
36. (A) (B) (C) (D)
37. (A) (B) (C) (D)
38. use line provided
39. (A) (B) (C) (D)
40. (A) (B) (C) (D)
41. (A) (B) (C) (D)
42. (A) (B) (C) (D)

Know It All! **Practice Test**

1. Spain conducts a lottery known as El Gordo, which means "The Fat One." It is one of the world's richest lotteries, though the winnings are spread among many. Assume that a jackpot to be split is billions of dollars. How would you write 1.7×10^9 in standard form?

   ____

2. The table below shows the number of people believed to have been aboard some of the ships that have disappeared in the Bermuda Triangle. What is the approximate number of people who have disappeared?

Ship or Plane	Number of People
Atlanta	290
Flight 19	14
U.S.S. Cyclops	309
Stavenger	43
Anglo-Australian	39
U.S.S. Scorpion	99

   A 790
   B 750
   C 840
   D 820

3. The capybara is the world's largest rodent. Think of a humongous guinea pig. A capybara can weigh 45 kilograms. What is the weight of a guinea pig if it weighs 3% of the weight of a capybara?

   A 0.5 kilograms
   B 1.35 kilograms
   C 1.65 kilograms
   D 2.5 kilograms

4. Some of the gladiators in ancient Rome were women. The practice was widespread until it was outlawed in the third century B.C. Look at the two women below and examine the angles between their bodies and their swords. Are the figures similar or congruent? Why?

____

5. The cartoon characters the Simpsons first appeared as animated shorts on *The Tracy Ullman Show* in 1989. Some of the awards the creators of *The Simpsons* television series won in the show's first 10 seasons included 15 Emmy awards, 9 International Animated Film Society awards, and 3 Environmental Media Awards. Which of the following shows the commutative property for the total awards the show won in its first 10 seasons?

   A $(15 + 9)3 = (3 \times 9) + (3 \times 15)$
   B $15 + 9 + 3 = 3 + 15 + 9$
   C $15 + (9 + 3) = (15 + 9) + 3$
   D $15(9 + 3) = 15(9) + 15(3)$

What's the Heat Index?

You may have heard a weather report that talks about the heat index, especially on hot summer days. The heat index is what the combination of the actual temperature and the relative humidity feels like on your skin. So on a hot day, it might actually feel hotter than it really is!

6. The numbers below show the heat index on consecutive days from August 4 through 12.

   105, 93, 102, 83, 108, 105, 111, 99, 110

   What is the range of the heat indexes above?

   ____

7. There are more than 6,000 languages spoken in the world, many of which are spoken by the incredibly diverse population of New York City. On any given day, the 7 train in New York City takes thousands of immigrants from their homes in Queens, New York, to work in Manhattan. A student polled 100 people in a subway car on their native language. Based on the circle graph below, what native language is spoken by the most people who took the survey?

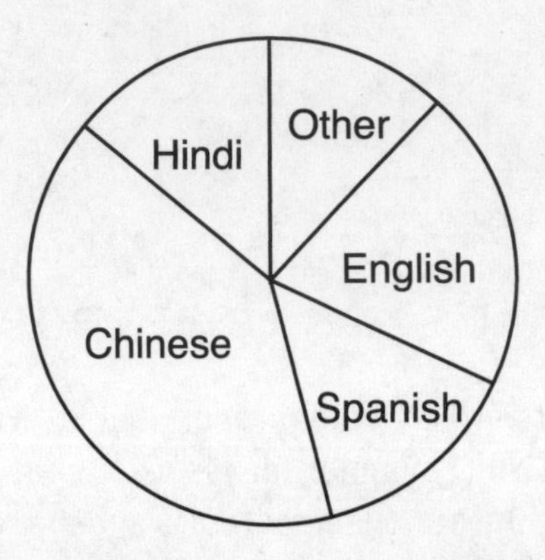

A Hindi
B Spanish
C Chinese
D English

8. Andrea wants to create a sculpture out of colored marbles. There are 3 blue marbles and 3 red marbles in a cup. What's the probability that after picking twice, she will have picked 2 blue marbles?

A $\frac{5}{11}$
B $\frac{1}{5}$
C $\frac{1}{6}$
D $\frac{1}{2}$

The Flea Circus

Did you know that fleas perform? Yep, in around 1600, the flea circus was invented. Doesn't that make you itch? There are still flea circuses today, in which fleas are trained to dance, pull coaches, perform high-wire acts, and be shot out of a cannon.

9. A flea can pull 160,000 times its own weight, which is about the same as an average adult pulling 12,000 tons. What is an equivalent of 12,000 tons?

   A 6 pounds
   B 24 pounds
   C 24,000,000 pounds
   D 192,000 ounces

10. Orlando and Shem just bought cell phones. Orlando has a plan of $34.99 per month for 350 minutes and 7 cents for every minute over that. Shem has a plan of $39.99 per month for 300 minutes and 5 cents for every minute over that. Who will pay less to talk for 500 minutes in one month? Explain how you got your answer.

____

____

____

11. One of the competitive events in the X Games is called the bike stunt vert. Several competitors have thrilled crowds by completing a 900. A 900 is 2.5 revolutions on a bike in the air. Rider Mat Hoffman was the first to do it no-handed, in August 2002. Amazing! Place the following fractions and decimals on the number line below and label them.

$2.5 \quad 1\frac{1}{8} \quad 2\frac{1}{4} \quad 1.75$

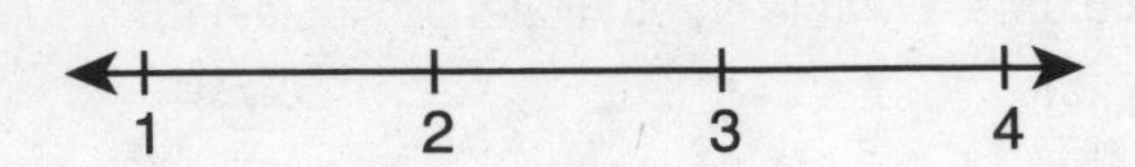

12. Hilary went to a bookstore and saw copies of a year-old edition of *The Guinness Book of World Records* on sale for \$3.75. She decided to buy 4 copies of the book to give as gifts. Which of the following uses the distributive property to find the total cost of the books?

A $\$3.75 + \$3.75 + \$3.75 + \$3.75 = \$3.75 + \$3.75 + \$3.75 + \$3.75$
B $4 \times \$3 + \$0.75 = \$0.75 + 4 + \$3$
C $4 + (\$3 - \$0.75) = \$3 - \$0.75 + 4$
D $4 \times (\$3 + \$0.75) = 4(\$3) + 4(\$0.75)$

13. Allana earns $22 per week baby-sitting and has $110 in savings. Assuming she doesn't spend any of her earnings, which equation shows how many weeks it will take for her to save $374?

A $22x + 110 = 374$
B $110x + 22 = 374$
C $22(x + 110) = 374$
D $110(22x) = 374$

14. One of the loudest burps ever recorded was 118.1 decibels by Paul Hunn on April 5, 2000. That's as loud a jackhammer! He and other contestants actually competed on British television. The scores of a belching contest are shown in the table below.

Contestant	Jack	Ginger	Byron	Seth
Decibels	115.3	108.6	112.4	113.7

What is the mean score of the belching contest?

A 113.3
B 112.2
C 112.5
D 112.9

15. A figure has four sides of equal length, its diagonals are perpendicular, and two of its angles measure 70°. What is it?

 A a rectangle
 B a square
 C a pentagon
 D a rhombus

16. The brain has about 100 billion brain cells. A signal from one brain neuron to another travels about 200 miles per hour. Which of the following are factors of 200?

 A 2, 4, 5, 6, 10
 B 2, 3, 4, 5, 8
 C 2, 4, 5, 10, 12
 D 2, 4, 5, 8, 10

17. Here's a delicious after school snack for you. Mix together $1\frac{1}{2}$ cups of fish eyes, $\frac{2}{3}$ cup of hissing cockroach guts, $1\frac{3}{4}$ cups regurgitated worms, $2\frac{1}{3}$ cups raw crocodile liver, and $1\frac{1}{4}$ cups millipede legs. Before you bake it, sprinkle it with a little belly button lint. Lip-smacking good! Add the ingredients together to find out how much you'll have for you and your friends.

 A $6\frac{2}{3}$ cups
 B $7\frac{1}{4}$ cups
 C $6\frac{3}{4}$ cups
 D $7\frac{1}{2}$ cups

The Death of the Sun

The Sun is a star that gives us energy in the form of light and heat. As it gets older and begins to die, the Sun won't suddenly go out. Instead, it will gradually get hotter and larger and increase its energy level. In about five billion years, life on Earth will end because the Sun will have made it too hot to survive here.

18. The dramatic storms on the Sun are called flares. They are very violent, but they last for only a minute or so. The flashes of gases may have a temperature of up to 20,000,000 degrees Celsius. What is 20,000,000 in scientific notation?

    A 2×10^7
    B 2×10^5
    C 2×10^6
    D 2×10^8

19. The Sun produces an amazing amount of energy. In 1 minute, it could melt a layer of ice 11 miles thick! Fortunately for us, the energy level is much lower by the time it reaches Earth. What is about the same measurement as 11 miles?

    A 18 kilometers
    B 18 meters
    C 18 yards
    D 18 feet

20. A video game features an alien that moves along a path equal to $y = 3x - 4$. If the alien's path was graphed, which of the following ordered pairs would satisfy the graph?

A (0, –3)
B (4, 10)
C (2, –2)
D (–3, –13)

21. Look at the figure below. Draw a translation of the figure on the grid.

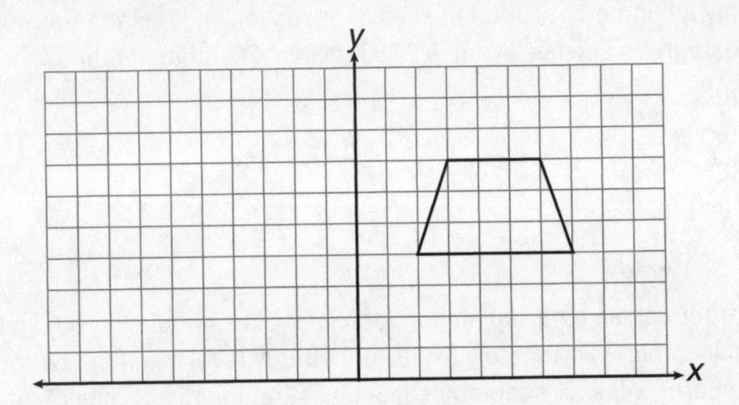

22. On May 25, 1999, Dominic Swaffer completed 1.4 baranis per second on a trampoline. A barani is a front somersault with a half twist. Wasn't he dizzy? Which property is used by the equation below to find the number of baranis that were completed in one minute?

$$1.4(60) = 60 \times 1.4$$

____

Wanted: Bat Nets

Colonel Percy Fawcett was hired to survey Bolivia for the Royal Geographic Society in 1906. He found many hazards in the jungle, including vampire bats. Though his team used mosquito nets at night to protect their sleeping bodies from the bites of insects, the nets were far from batproof! If any skin was near the net, a vampire bat was able to make a puncture wound and then lap up a person's blood. Sometimes in the morning, the nets would be covered with blood. Totally gross!

23. If a rectangular mosquito net had a width of m and a length of $m + 8$, which expression would represent the area of the rectangle?

    A $2m + 2(m + 8)$
    B $m + (m + 8)$
    C $m(m + 8)$
    D $2m(m + 8)$

24. Colonel Fawcett believed there was a lost ancient city in Brazil, which he called Z. Fawcett and his son went to find it, but they disappeared into the jungle and were never seen again. Look at the map below. The distance between the camp and the lost city is $2\frac{1}{2}$ inches on a ruler. What is the actual number of miles?

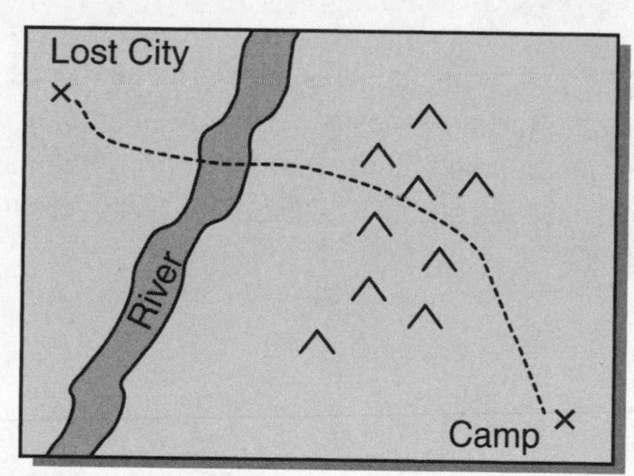

Scale
1 inch = 24 miles

    A 72 miles
    B 60 miles
    C 36 miles
    D 42 miles

Express Yourself

Have you noticed how many bumper stickers there are on cars these days? Bumper stickers are a very popular way for people to express themselves. They have become a way for people to share their interests, their values, their senses of humor, and the places they've visited.

25. Suppose you needed to design a bumper sticker for a school fund-raiser. You want it to be slightly larger than the sticker that was designed last year. The two bumper stickers below are similar. What is the length of the missing side of the larger bumper sticker?

14 cm

Barfdale
Middle School
Class of 2004

26 cm

x

39 cm

A 21 centimeters
B 18 centimeters
C 24 centimeters
D 23 centimeters

26. What male animal gets pregnant and gives birth? The seahorse. The male seahorse has a pouch in which the female deposits her eggs. The males carry them and give birth in 21 days. Over the course of three months, a group of male seahorses delivered 1,482 babies, 397 babies, and 1,205 babies. What is the estimate of the total number of babies?

A 4,000
B 3,100
C 3,000
D 4,100

27. Around the world, people have built walls for many centuries and for many reasons. Sometimes they are for the defense of a city or town, and sometimes they are a tribute to a leader or person of honor. On a separate sheet of paper, draw a bar graph that displays the data below about some notable walls. Don't forget to label the graph.

Great Wall of China . . . .25 feet high	Vietnam Veterans Memorial Wall . . . . . .10 feet high
Berlin Wall . . . . .13 feet high	Hadrian's Wall . . . . .14 feet high

28. A cuddly puppy is so sweet and so much fun to play with. As is the case with grown dogs, they come in all sizes. What would be an appropriate weight for a 12-week-old Labrador retriever puppy?

A 15 ounces
B 15 liters
C 15 grams
D 15 kilograms

SOS!

Have you ever heard the term SOS? SOS is a signal sent by ships to indicate that there is an emergency. However, the letters SOS are not an abbreviation for anything. In 1906, radio operators agreed that SOS would be the best distress signal because people interpreting messages could not misunderstand the 3 dots, 3 dashes, and 3 dots used to signal the letters.

29. What number comes next in the sequence below?

3, 6, 12, 24, ____

____

Can You Spare a Rhode Island?

In 1999, the United States Mint began circulating a series of quarters to commemorate each state in the Union. The 50 state quarters will be released in the order the states became part of the United States until the year 2008.

30. Juma has 5 Indiana, 5 New Jersey, 10 Alabama, 5 Maryland, and 15 Vermont quarters in a bag. If he picks a coin from the bag without looking, what is the probability that it will be an Alabama quarter?

    A 25%
    B 40%
    C 50%
    D 13%

31. In ancient times, salt was extremely valuable, even more valuable than gold. There are parts of the world where salt is still used as currency. If a bar of salt measures 20 centimeters long, 8 centimeters wide, and 6 centimeters high, what is its surface area?

    A 528 square centimeters
    B 600 square centimeters
    C 656 square centimeters
    D 578 square centimeters

32. The black mamba may be the deadliest snake in the world. It's mean, it's large, and it carries deadly venom. It gets its name from the black lining of its mouth. The lengths of four black mambas are given below. Which one is the longest?

Annie $8\frac{3}{8}$ feet

Bosco $8\frac{1}{4}$ feet

Calvin $8\frac{3}{6}$ feet

Dmitri $8\frac{3}{10}$ feet

A Annie
B Bosco
C Calvin
D Dmitri

33. A drawing of one of the largest black mambas on record is 9 centimeters long. If the scale of the drawing is 1 centimeter = 0.5 meters, how long is the actual snake?

A 45 meters
B 40 meters
C 4.5 meters
D 4.0 meters

34. A black mamba can be exceptionally fast on the ground. Its speed in miles per hour equals the whole number between $\sqrt{45}$ and $\sqrt{59}$. What is its speed?

A 9 miles per hour
B 7 miles per hour
C 6 miles per hour
D 8 miles per hour

35. Plastic was invented in 1907 by mixing the chemicals phenol and formaldehyde and heating them. Plastic was originally called Bakelite and was used to make radio cabinets, telephones, distributor caps, pot handles, and many other items. Look at the pot handles below. Which term best describes the figure on the right?

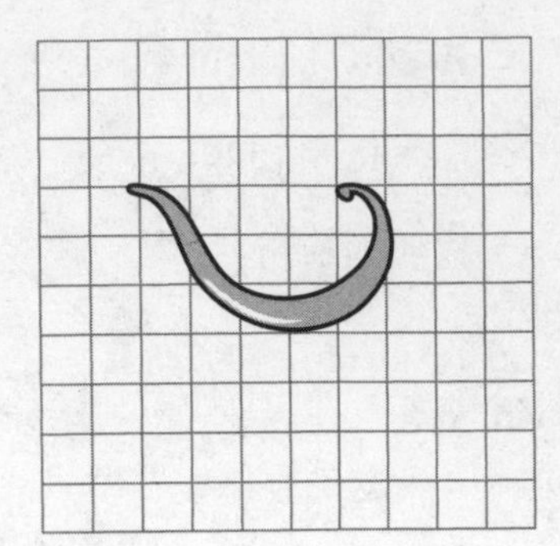

A reflection
B translation
C rotation
D flip

36. A black hole forms when a massive star collapses on itself. Black holes allow no light to escape, and they are very small and extremely tightly packed. Albert Einstein had a theory that black holes distort space and time to form a shape like a funnel. The opening of a black hole funnel may have a diameter of 18 miles. If so, what would the hole's circumference be?

A 58.82 miles
B 113.04 miles
C 56.52 miles
D 115.44 miles

37. Tamara made a chocolate ice cream roll that is shaped like a zombie's leg and is 14 inches long. She wants to cut a portion 2 inches wide and another portion 3 inches wide to give to her cousins. She'll divide the rest into $1\frac{1}{2}$-inch portions. How many pieces will Tamara cut in all?

A 8
B 7
C 5
D 6

38. Groups of animals all have different names, such as a pod of whales. Here are some more interesting ones: an army of frogs, a clowder of cats, a leap of leopards, a murder of crows, a bask of crocodiles, and a drift of hogs. Look at the crow below and measure the angle formed by the bird and the ground as indicated.

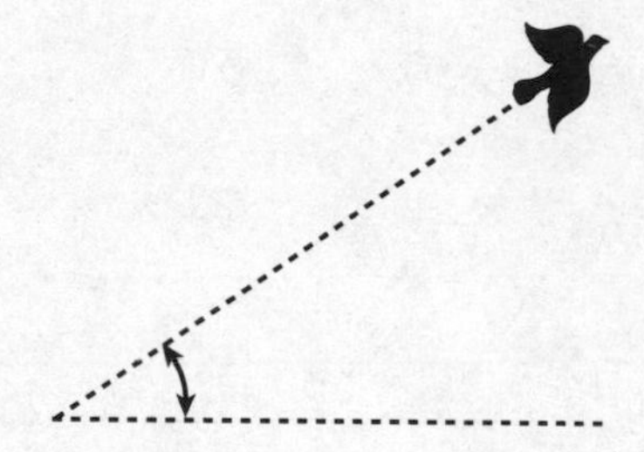

____

39. When Miles and Cleo went on vacation, they photographed a solid figure. The figure has faces that each have three sides and angles that add up to 180°. The base is a square. What is this solid?

A a cylinder
B a cone
C a triangular prism
D a pyramid

40. Have you heard the saying "blonds have more fun"? Well, maybe it's because they have more hair. The average blond has 140,000 hairs on her head while the average redhead has 90,000 hairs. No matter how much hair a person has, some is growing while some is resting as it gets ready to fall out. If 126,000 hairs on the average blond are growing, what percent of the hair is growing?

A 10%
B 90%
C 71%
D 15%

41. Do you like peanut butter? Well, about 75% of the households in the United States do. It's a good source of protein and has many vitamins and minerals. Peanut butter is used in many recipes, including some soups. If a jar of peanut butter is 8 inches high and has a radius of 5 inches, what is its volume?

A 126 cubic inches
B 212 cubic inches
C 628 cubic inches
D 653 cubic inches

42. The world's largest snake is the anaconda, which can grow up to 30 feet long. If a 30-foot anaconda weighs 400 pounds, how much would one that's 18 feet long weigh if it had the same ratio of length to weight as the larger anaconda?

A 200 pounds
B 240 pounds
C 180 pounds
D 220 pounds

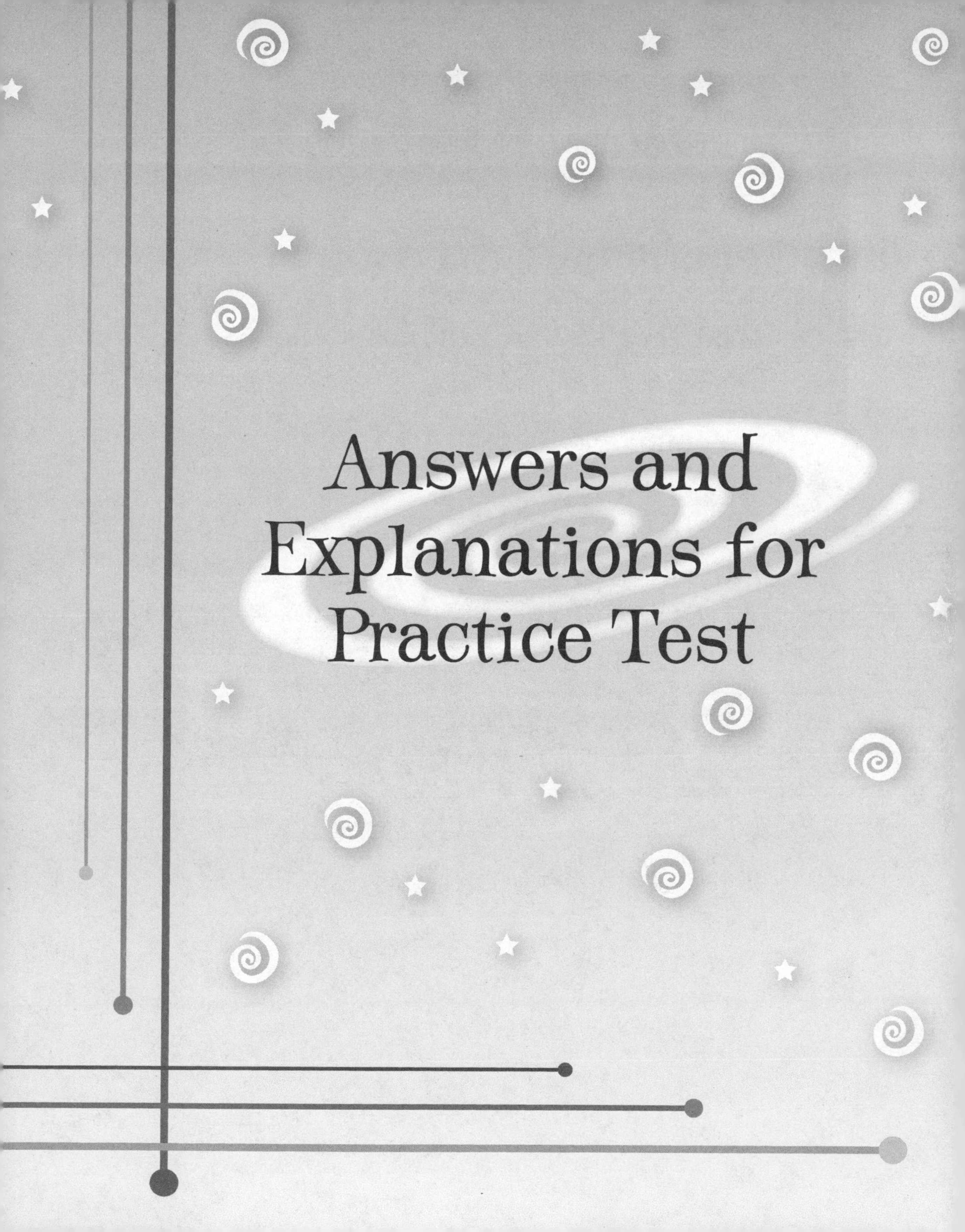

Answers and Explanations for Practice Test

Answers and Explanations for the Practice Test

1. The correct answer is 1,700,000,000. Remember that the exponent in 1.7×10^9 tells you how many places there are from the decimal point to the end of the number. In this case, the exponent is 9 so you count the 7 as one place and add eight zeros. If you need help with exponents, review chapter 7.

2. **A** Answer choice (A), 790, is correct. You're asked for the approximate number of people, which means that you should round each of the numbers in the second column. The number 290 can remain the same, 14 rounds to 10, 309 rounds to 310, 43 rounds to 40, 39 rounds to 40, and 99 rounds to 100. Add the rounded numbers:

$$290 + 10 + 310 + 40 + 40 + 100 = 790.$$

3. **B** Answer choice (B), 1.35 kilograms, is correct. The capybara weighs 45 kilograms, and the guinea pig's weight is only 3% of that. Use multiplication to find the answer. $45 \times 0.03 = 1.35$. (Be careful that you put the decimal point in the correct place.) So the cute little guinea pig weighs 1.35 kilograms, a lot less than that humongous capybara!

4. This is an open-response question. You would need to give the correct answer and an explanation to get full credit. If you had only written "similar," you would have had points deducted because you hadn't given a reason for your answer. Make sure that you answer all parts of an open-response question. The figures are similar. Look carefully at the two gladiators. One is larger than the other. Even though the angle between body and sword is the same for both—about 75°—the size of each is different. Remember the tip in chapter 14 that said similar figures will match up if one of them is magnified or shrunk? A good response could have been "The two figures are similar because their angles are the same but their size is different."

5. **B** Answer choice (B), $15 + 9 + 3 = 3 + 15 + 9$, is correct. Remember that the commutative property states that $3 + 4 = 4 + 3$ or $3 \times 4 = 4 \times 3$. If you aren't sure which of the answers is correct, use Process of Elimination. Answer choice (A) illustrates the distributive property, which isn't correct. Eliminate it. Answer choice (C) illustrates the associative property. Eliminate it. Answer choice (D) illustrates the distributive property again. Answer choice (B) is correct.

6. The range is 28. To find the range of a set of numbers, first put the numbers in numerical order from least to greatest.

   83, 93, 99, 102, 105, 105, 108, 110, 111

   The range is the difference between the greatest and least numbers. $111 - 83 = 28$. If you want to review how to find mean, median, mode, and range, see chapter 18.

7. **C** Answer choice (C), Chinese, is the correct answer. The question asked which language is spoken by the most people in the subway car. The circle graph only shows the top four languages, but from that you can determine which is spoken the most. You just need to find the one with the largest pie piece. Chinese takes almost half of the circle graph, and the other four take up the rest. That means that Chinese is the native language spoken by the most people in the subway car.

8. **B** Answer choice (B), $\frac{1}{5}$, is the correct answer. The cup contains 3 blue marbles and 3 red marbles. That's a total of 6 marbles. The probability of picking blue the first time is 3 out of 6, or $\frac{1}{2}$. Now, there are 5 marbles in the cup, and 2 are blue. The probability of picking a blue marble the second time is 2 out of 5, or $\frac{2}{5}$. To find the probability of picking 2 blue marbles in consecutive picks, multiply the two probabilities together. $\frac{1}{2} \times \frac{2}{5} = \frac{2}{10} = \frac{1}{5}$.

9. **C** Answer choice (C), 24,000,000 pounds, is the correct answer. Okay, no more itchy fleas. Instead, you need to determine which weight is equivalent to 12,000 tons. You might not know what 12,000 tons is actually like, but you can use Process of Elimination to find the correct answer. Answer choice (A) is 6 pounds. That's about the weight of two schoolbooks. Nowhere near 12,000 tons! Eliminate it. Choice (B), 24 pounds, is the weight of a little baby (like your pesky brother or sister). Eliminate it. Answer choice (D), 192,000 ounces, is the same as 12,000 pounds, not 12,000 tons. Eliminate it. 24,000,000 pounds is the correct answer. By the way, 12,000 tons is about the weight of 2,000 grown elephants.

10. Orlando's plan will cost less. To find the correct answer, you need to find the cost of 500 minutes for each plan. Orlando's plan is $34.99 plus 7 cents for every minute over 350 minutes. You could use this equation: $\$34.99 + \$0.07(500 - 350) =$ cost. The cost is $45.49. Shem's plan is $39.99 plus 5 cents for every minute over 300 minutes. You could use this equation: $\$39.99 + \$0.05(500 - 300) =$ cost. The cost is $49.99. A good answer would be as follows: "Orlando will pay less for 500 minutes because his plan would cost $45.49. Shem's plan would cost $49.99. I determined the answer by writing an equation for each phone plan." Showing your work for an open-response question is a good idea. That way, if you make a mistake in your arithmetic, you will still receive partial credit. As long as you show that you know how to do it, you're likely to receive some credit.

11. The correct number line is shown below.

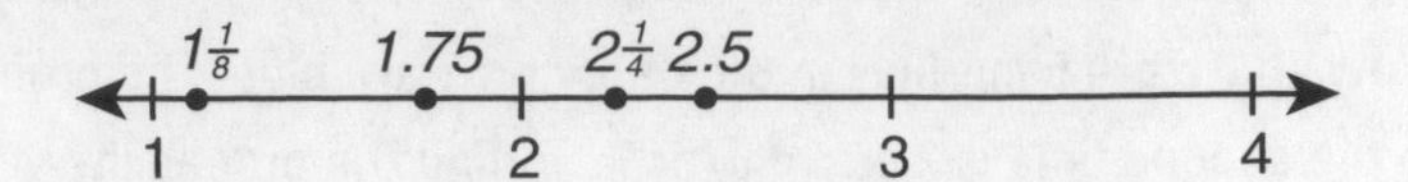

To make this easier for yourself, convert the fractions to decimals or the decimals to fractions. $2.5 = 2\frac{1}{2}$, $1\frac{1}{8} = 1.125$, $2\frac{1}{4} = 2.25$, and $1.75 = 1\frac{3}{4}$. Now, it's easier to compare the numbers and place them correctly on the number line.

12. **D** Answer choice (D), 4 × ($3 + $0.75) = 4($3) + 4($0.75), is the correct answer. The question wants you to find the answer choice that uses the distributive property. The distributive property is the one that distributes the number you multiply by to the other numbers. If you can't find the answer, use Process of Elimination. Answer choice (A) is $3.75 + $3.75 + $3.75 + $3.75 = $3.75 + $3.75 + $3.75 + $3.75. There's nothing being distributed there. Eliminate it. Answer choice (B) is 4 × $3 + $0.75 = $0.75 + 4 + $3. Here, the number 4 should be multiplied by $3.75. Eliminate it. Answer choice (C) is 4 + ($3 − $0.75) = $3 − $0.75 + 4. That uses subtraction, and the distributive property only works for addition and multiplication. Answer choice (D) is 4 × ($3 + $0.75) = 4 ($3) + 4($0.75), and that is distributing. The last choice is correct.

13. **A** Answer choice (A), $22x + 110 = 374$, is the correct answer. The question wants to know the number of weeks—make that x—that it will take for Allana to earn $374. Look at each equation carefully to determine what it does. Answer choice (A) multiplies 22 (the amount she makes each week) by x and adds 110 (the amount Allana already has) to get 374 (the amount she wants to earn). This sounds right. In choice (B), the number of weeks is multiplied by 110. That's not right, because she doesn't earn $110 a week. In answer choice (C), the amount she earns each week is multiplied by the number of weeks *and* the amount she already has. That's definitely not right. In answer choice (D), the $110 in the bank is multiplied instead of added. That's not right. Answer choice (A) is correct.

14. **C** The mean score is 112.5. The question asks you to find the mean, or average, of the set of numbers. Take the numbers in the decibel row and add them together.

$$115.3 + 108.6 + 112.4 + 113.7 = 450$$

Now, divide by the number of items in the set, which is 4.

$$450 \div 4 = 112.5$$

The correct answer is 112.5, answer choice (C).

15. **D** The figure is a rhombus. Read the description step by step to find the right answer. There are four sides that are equal. That's a square or a rhombus. The diagonals are perpendicular. Not much help there; it's still a square or a rhombus. Two of the angles are 70°. Ahh, there's a good clue. The angles of a square are all 90°. This cannot be a square; it must be a rhombus, answer choice (D). Review chapter 11 if you need help with polygons or solids.

16. **D** Answer choice (D), 2, 4, 5, 8, 10, are the factors. Do you ever feel after a long, hard day that your brain neurons are firing at about 5 miles per hour? Hopefully, they're firing fine now, so that you can determine the factors of 200. Some factors are easy to determine. You know that any even number is divisible by 2. You also know that a number that ends in a zero is divisible by 5 and 10. Now you have 2, 5, and 10. 4 is another factor, because $200 \div 4 = 50$. For a number to be evenly divisible by 3, the digits must add up to a number divisible by 3. 200 is not evenly divisible by 3, because $2 + 0 + 0 = 2$. For a number to be evenly divisible by 6, it must be divisible by 2 and 3. 200 isn't divisible by 3, so it's not divisible by 6. The factors so far are 2, 4, 5, and 10. What about 8? Yes, 200 is divisible by 8. But 200 is not evenly divisible by 12. The only answer choice that lists numbers that are all factors of 200 is (D), 2, 4, 5, 8, and 10.

17. **D** Didn't this question make you hungry? To add the measurements together, you'll need to change the fractions to have a common denominator. The lowest one in this case is 12. $1\frac{1}{2} = 1\frac{6}{12}$, $\frac{2}{3} = \frac{8}{12}$, $1\frac{3}{4} = 1\frac{9}{12}$, $2\frac{1}{3} = 2\frac{4}{12}$, and $1\frac{1}{4} = 1\frac{3}{12}$. Now, add them together.

$$1\frac{6}{12} + \frac{8}{12} + 1\frac{9}{12} + 2\frac{4}{12} + 1\frac{3}{12} = 5\frac{30}{12} = 7\frac{6}{12} = 7\frac{1}{2}$$

There will be $7\frac{1}{2}$ cups of yummy, melt-in-your-mouth snacks. Don't forget the belly button lint! Answer choice (D), $7\frac{1}{2}$ cups, is correct.

18. **A** In scientific notation, 20,000,000 is 2×10^7. To find the correct exponent when converting to scientific notation, count the number of places from the 2 to the end of the last zero. There are 7 places, so the exponent will be 7. Multiply the 2 by 10 to the seventh power, or 2×10^7.

19. **A** Choice (A), 18 kilometers, is the correct answer. 11 miles is a long distance. A kilometer is another long distance that's a bit more than half a mile. This looks like a good answer. You can use Process of Elimination just to make sure. Answer choice (B), 18 meters, is about the height of a 5-story building. That's not even 1 mile. Eliminate it. Answer choices (C), 18 yards, and (D), 18 feet, are even less of a distance than 18 meters. They can't be right, either. 18 kilometers is correct.

20. **D** A video game with an alien? Cool! Unfortunately, that alien can't solve this problem for you. You'll have to plug each pair of numbers into the equation $y = 3x - 4$ to see which one works. The first answer choice would make the equation $-3 = 3(0) - 4$, or $-3 = -4$. Nope, that's not right. Eliminate it. The second choice is $10 = 3(4) - 4$, or $10 = 8$. That's wrong. The third choice is $-2 = 3(2) - 4$, or $-2 = 2$. That doesn't work. Last one (should be right): $-13 = 3(-3) - 4$, or $-13 = -13$. Yee-ha! That's the right one! Answer choice (D), (–3, –13), is the correct answer.

21.

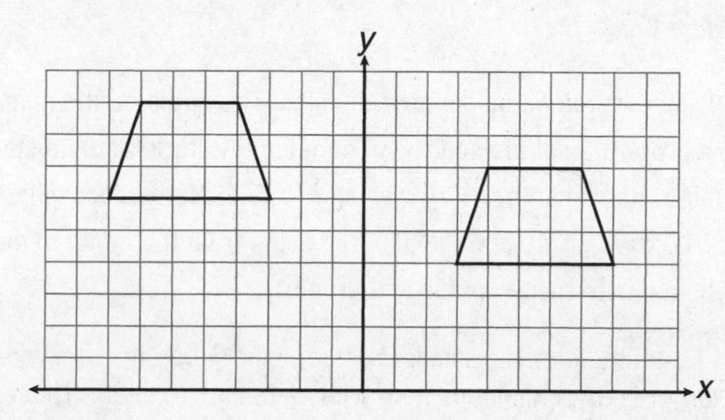

A translation is a change in position of a figure. It is also called a slide. The trapezoid can be moved in any direction on the grid, as long as it isn't turned in any way. If you'd like more help with transformations, see chapter 15.

22. The commutative property is the correct answer. Remember that the commutative property is the one that switches the order of numbers that are added or multiplied and gets the same answer. In this case, 1.4(60) would be the same as 60×1.4. If you want more practice with the properties of numbers, review chapter 8.

23. **C** Answer choice (C), $m(m + 8)$, represents the area of the mosquito net. Remember that the formula for the area of a rectangle is $A = lw$. Use $m + 8$ for the length (l) and m for the width (w). This would give you $(m + 8) \times m$, which is (because of the commutative property) the same as $m(m + 8)$.

24. **B** Answer choice (B), 60 miles, is the actual number of miles. The scale on the map is 1 inch = 24 miles. Set up a proportion of the scale $\left(\frac{1}{24}\right)$ equaling the ratio of the distance measured on the map to the actual distance $\left(\frac{2.5}{x}\right)$, that is, $\frac{1}{24} = \frac{2.5}{x}$. Now, cross multiply (and, technically, divide by 1). $x = 2.5(24)$, or $x = 60$. It's 60 miles from the camp to the lost city. Look at chapter 5 if you'd like to review ratios and proportions.

25. **A** The missing width is 21 centimeters. The smaller rectangle is 14 centimeters wide by 26 centimeters long. The larger is 39 centimeters long with an unknown width. These are similar rectangles because they match in shape but are different sizes. You can use a proportion to find the missing width.

$$\frac{14}{x} = \frac{26}{39}$$

$$26x = 14 \times 39$$

$$x = \frac{546}{26}$$

$$x = 21$$

The width of the larger rectangle is 21 centimeters. Choice (A) is correct.

26. **B** Answer choice (B), 3,100, is the approximate total of baby seahorses. You know that when you estimate, you round numbers. 1,482 is rounded to 1,500, 397 is rounded to 400, and 1,205 is rounded to 1,200. Add them together: 1,500 + 400 + 1,200 = 3,100.

27. A bar graph as below would be best to display this information.

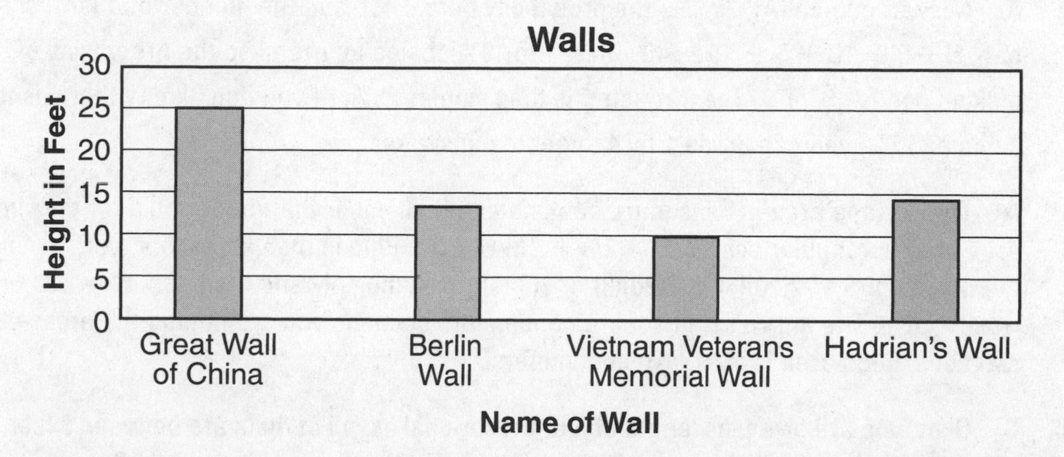

You would have received full credit if you had completed a graph like the one above. If you hadn't titled it or labeled the axes, you would have received partial credit. Remember that when you are answering open-response questions, you should be as complete as you can. You can lose credit if you don't give all the information.

28. **D** Even if you don't know the weight of a Labrador retriever puppy, you can use Process of Elimination to narrow down your choices. Answer (A) is 15 ounces. That isn't even a pound. A 12-week old puppy with floppy ears and big feet has got to weigh more than that! Choice (B) is 15 liters. Liters are a measurement of liquid, so that can't be right. Answer choice (C) is 15 grams. That's about the same as 15 paper clips! Answer choice (D), however, is 15 kilograms. Hmmm, that sounds about right: big floppy ears, a long wagging tail, big feet, and a round belly. Yep, answer choice (D), 15 kilograms, is the correct answer.

29. The next number in the sequence is 48. The difference between 3 and 6 is 3. The difference between 6 and 12 is 6. (Are you starting to see a pattern?) The difference between 12 and 24 is 12. Look at that: each difference has been doubled. So the next number would be twice 12, which is 24. Add 24 to 24 and you get 48.

30. **A** Answer choice (A), 25%, is the probability here. First, add the number of quarters in the bag. $5 + 5 + 10 + 5 + 15 = 40$. There are 10 Alabama quarters, so the probability of picking one is $\frac{10}{40}$, or $\frac{1}{4}$. The fraction $\frac{1}{4}$ is the same as 25%. (If you don't know that $\frac{1}{4}$ is the same as 25%, you can divide 1 by 4 to get the answer.)

31. **A** The surface area is 656 square centimeters. Remember that the formula for the surface area of a rectangular solid is $S = 2lh + 2lw + 2wh$. Plug in the values that were given in the question: $S = 2(20)(6) + 2(20)(8) + 2(8)(6)$. Now, multiply and then add. $S = 240 + 320 + 96 = 656$. The answer is in square centimeters because you are adding the areas of the six faces. Each area is in square centimeters.

32. **C** Okay, you are given the length of four creepy snakes. All of them are between 8 feet and 9 feet long. (8 feet? Can you imagine a snake that long, especially one that's mean?) All you need to concern yourself with are the fractions and putting them in the correct order.

The first is $\frac{3}{8}$, and it's larger than the second, $\frac{1}{4}$, because $\frac{1}{4} = \frac{2}{8}$. So far, $\frac{3}{8}$ is the largest. The next fraction is $\frac{3}{6}$, which is the same as $\frac{1}{2}$. That's larger than $\frac{3}{8}$, because $\frac{4}{8} = \frac{1}{2}$. The last one is $\frac{3}{10}$, and that's smaller than $\frac{3}{6}$, because $\frac{5}{10} = \frac{1}{2}$. The largest fraction is $\frac{3}{6}$, so the largest snake is the big, scary black mamba that is $8\frac{3}{6}$ feet long, which is Calvin, answer choice (C).

33. **C** The length of the actual snake is 4.5 meters. The drawing shows a black mamba that is 9 centimeters. The scale is 1:0.5. Set up a proportion to find the actual length. $\frac{1}{0.5} = \frac{9}{x}$. Solve for x by cross multiplying and dividing. $x = 4.5$. The actual snake is 4.5 meters long. Look at the answer choices. The first one of 45 meters might make you rethink your answer, but if you've taken care to place the decimal point correctly, you will know that 4.5 is correct. Besides, can you even imagine a black mamba that is 45 meters long? It would have to be in a horror movie! Answer choices (B), 40 meters, and (D), 4.0 meters, can't be the answer to the multiplication problem, never mind the decimal point. Answer choice (C) is correct.

34. **B** The speed of a black mamba is 7 miles per hour. Oh, great! They're long, they're mean, and now they're fast. Just the kind of snake to avoid! To find the answer to the question, you need to find the whole number that lies between $\sqrt{45}$ and $\sqrt{59}$. Think of squares of whole numbers to find the answer. The square of 6 is 36, which is too low. The square of 7 is 49. That looks like the right answer. Try one number higher. The square of 8 is 64, which is too high. The correct answer is 7.

35. **C** Rotation is the correct answer. If you're not sure of the answer, use Process of Elimination. Answer choice (A) is reflection, but a reflection is a figure that has been flipped. The figure in the graph isn't flipped. Eliminate the first choice. Choice (B) is translation, but that's a figure that has slid in one direction. This figure hasn't slid, so eliminate that answer. Answer choice (C) is rotation, which is a figure that has been turned. That's what's been done to the figure on the graph. That's the right answer. Answer choice (D) is the same as the first answer choice: flip = reflection.

36. **C** Choice (C), 56.52 miles, is the correct circumference. The formula for the circumference of a circle is $C = \pi d$. Plug in the diameter you've been given. $C = \pi 18$. Use 3.14 for π. $C = (3.14)18 = 56.52$ miles, which is the third answer choice.

37. **A** The correct answer is 8. The ice cream zombie leg is 14 inches. (Don't you hope there are sprinkles on that?) Two pieces are cut that are 2 inches and 3 inches. $14 - 5 = 9$. 9 inches are left. That portion will be divided into $1\frac{1}{2}$-inch pieces. $9 \div 1\frac{1}{2} = 6$ pieces. So there are 6 pieces *plus* the 2 original ones, which equals 8.

38. The correct measure is 35°. To find the measure of this angle, put your protractor base on the bottom line (the ground) and find the angle of the crow's flight on your protractor. The angle of the crow's flight measures 35°.

39. **D** Read the description carefully. First, you're told that it's a solid figure, so you know there will be more than two dimensions. The figure's faces each have three sides and have angles that add up to 180°. That's a triangle. Next, you're told that the base is a square. So you have a square base with triangular faces. That's a pyramid. (A triangular prism has faces that are rectangles and bases that are triangles.) Answer choice (D), a pyramid, is correct.

40. **B** The correct answer is 90%. You need to find the percent of hair that is growing on the average blond head. 126,000 hairs are growing, and there are 140,000 hairs on the average blond. To find the percent, divide the growing hairs by the average hairs, or 126,000 by 140,000. That equals 0.9, which is the same as 90%.

41. **C** Choice (C), 628 cubic inches, is the volume. The formula for the volume of a cylinder is $V = \pi r^2 h$. Plug in the values you've been given. $V = \pi 5^2(8) = \pi 25(8)$. Use 3.14 for π. $V = 628$ cubic inches.

42. **B** Answer choice (B), 240 pounds, is correct. To find the answer to this question, you would use a proportion. Set the ratio of length to weight of the really humongous snake equal to that of the smaller snake. $\frac{30}{400} = \frac{18}{x}$. Now, cross multiply. $30x = 400(18)$. Divide both sides by 30 to find that $x = 240$. The "little" snake weighs in at 240 pounds, while the bigger snake weighs about as much as a sumo wrestler.

If students need to know it, it's in our Know It All! Guides!

Know It All!
Grades 3–5 Math
0-375-76375-9 • $14.95

Know It All!
Grades 3–5 Reading
0-375-76378-3 • $14.95

Know It All!
Grades 6–8 Math
0-375-76376-7 • $14.95

Know It All!
Grades 6–8 Reading
0-375-76379-1• $14.95

Know It All!
Grades 9–12 Math
0-375-76377-5 • $14.95

Know It All!
Grades 9–12 Reading
0-375-76374-0 • $14.95

Available at your local bookstore